Memoirs of the American Mathematical Society

Number 47

PERIODIC SOLUTIONS OF PERTURBED SECOND-ORDER AUTONOMOUS EQUATIONS

BY

W. S. LOUD

1964

AMERICAN MATHEMATICAL SOCIETY

190 Hope Street, Providence, Rhode Island

Library of Congress Catalog Number 52-42839

CONTENTS

PART I: PRELIMINARIES

1.1. Introduction. In the real autonomous second-order differential equation

$$x'' + g(x, x') = 0 \qquad (' = d/dt) \tag{1.1}$$

let $g(x, x')$ be sufficiently regular for values of x and x' considered. We shall always assume that $g \in C^2$, so that solutions of (1.1) will be uniquely determined by initial conditions. We suppose that (1.1) has a periodic solution, $x_0(t)$. The solution $x_0(t)$ may be either identically constant, or variable. If $x_0(t)$ is variable, let its least period be L_0.

Now suppose a periodic forcing term is added to the right member of (1.1)

$$x'' + g(x, x') = \varepsilon f(t, x, x', \varepsilon), \tag{1.2}$$

where $f(t, x, x', \varepsilon)$ is also real and sufficiently regular, and has least period $T > 0$ in t. (Thus f must depend explicitly on t.) We shall always assume that $f \in C^1$, so that solutions of (1.2) are uniquely determined by initial conditions. If $x_0(t)$ is not constant, we assume that L_0 is a rational multiple of T.

We shall seek periodic solutions of (1.2) for small ε of the form

$$x = x_0(t) + \varepsilon x_1(t) + o(\varepsilon) \tag{1.3}$$

Received by the editors June 26, 1961.

The research for this paper was supported in part by the Office of Ordnance Research, U. S. Army, Contract No. DA-11-022-ORD-1869. A major part of the paper was written while the author was a visiting member of the Mathematics Research Center, U. S. Army, at the University of Wisconsin, Contract No. DA-11-022-ORD-2059.

of period L, where L is the least common multiple of T and L_0 when $x_0(t)$ is variable, and $L = T$ when $x_0(t)$ is constant.

Remarks:

1. Since we are studying solutions of the form (1.3) for small ε, the hypotheses on g and f are needed only for (x, x') in some neighborhood of the orbit $(x_0(t), x_0'(t))$.

2. There is no loss in generality in assuming the period T to be independent of ε. If T depended (differentiably) on ε, an appropriate change of variable would bring the equation to the form (1.2) with t independent of ε.

The methods used in this paper will be implicit function theorem techniques. They were used by the author [8] for the equation

$$x'' + cx' + g(x) = \varepsilon f(t) \tag{1.4}$$

and in [9] for the vector system

$$x' = g(x) + \varepsilon f(t, x, \varepsilon). \tag{1.5}$$

In [8] (1.4) was considered first for $c = 0$. Then with ε found sufficiently small and fixed, c was chosen sufficiently small. One of the results obtained in the present paper shows how small c must be compared to ε.

It is possible to transform (1.2) to a special case of (1.5) with vectors of dimension 2. The paper [9] considers only the case that $x_0(t)$ is variable and is an isolated periodic solution, corresponding to the assumption that the related variational equation has only one characteristic multiplier equal to +1. In spite of the slight simplification afforded by this transformation we choose to consider the scalar equation (1.2) as it stands. We believe that second-order equations are of interest for their own sake, and not merely as a special case of more general systems.

Certain cases of (1.2) have been studied for a long period of time. The equation

$$x'' + g(x) = \varepsilon f(t, x, x', \varepsilon) \tag{1.6}$$

was studied by Kac [6] in the analytic case by power series methods. When

$g(x, x')$ is linear in x and x', (1.2) represents a nearly linear system, which has been considered very extensively. See the paper by Friedrichs and Stoker [4] and the book by Cesari [2], both of which contain large bibliographies.

1.2. <u>The Autonomous Equation and the Variation Equation</u>

The basic assumption of this paper is the existence of a periodic solution, $x_0(t)$, of equation (1.1). It is by no means always the case that such solutions exist. For example, the linear equation

$$x'' + Bx' = C$$

has no periodic solutions when $C \neq 0$. It is also possible that the only periodic solutions are constants. For example if $g(x)$ is strictly increasing, and takes both signs, and $c \neq 0$, the only periodic solution of

$$x'' + cx' + g(x) = 0$$

is the constant solution $x = a$, where a is the unique zero of $g(x)$. However, there are many significant cases in which (1.1) has nonconstant periodic solutions. Among them are the following:

(1) The Lienard equation

$$x'' + f(x)x' + g(x) = 0, \tag{1.7}$$

which has van der Pol's equation

$$x'' + \mu(x^2 - 1)x' + x = 0 \tag{1.8}$$

as a special case. Under appropriate conditions, (1.7) has an isolated nonconstant periodic solution. For an extensive discussion see Minorsky's book [10], Chapter VI.

(2) The equation

$$x'' + g(x) = 0, \tag{1.9}$$

where $g(x)$ is not linear, but $xg(x) > 0$ for $x \neq 0$ and $g(0) = 0$. A typical case is Duffings's equation where $g(x) = \alpha x + \beta x^3$. Every solution of (1.9) near $x \equiv 0$ is periodic. There is the constant solution $x \equiv 0$, while the remaining solutions are nonconstant, and have periods varying with amplitude as a rule. The variation of period with amplitude and some of its consequences are discussed in [8], Section 2.

(3) The undamped linear equation

$$x'' + k^2 x = 0. \tag{1.10}$$

Here all solutions are periodic, and all nonzero solutions have the same period, $2\pi/k$. This gives rise to the nearly linear case mentioned earlier.

To study perturbations of the periodic solution $x_0(t)$, it is necessary to consider the related variation equation

$$L(y) \equiv y'' + g_{x'}(x_0(t), x_0'(t))y' + g_x(x_0(t), x_0'(t))y = 0. \tag{1.11}$$

It turns out that the problem of existence of a periodic solution of (1.2) in the form (1.3) is simple or complicated according as (1.11) does not or does have a nontrivial solution of the same period as that sought for $x(t)$. This condition is derived in [3], Chapter 13, and will be much in evidence in the remainder of this paper.

If $x_0(t)$ is constant, equation (1.11) has constant coefficients. The only cases in which it has a nontrivial periodic solution of period T are (a) when $g_x(x_0, 0) = 0$, and (b) when $g_{x'}(x_0, 0) = 0$ and $g_x(x_0, 0) = 4n^2\pi^2T^{-2}$ for some integer n.

If $x_0(t)$ is nonconstant, equation (1.11) has periodic, usually variable, coefficients. There is always the nontrivial periodic solution $y = x_0'(t)$. To see this, differentiate the identity

$$x_0''(t) + g(x_0(t), x_0'(t)) \equiv 0$$

with respect to t. The result is $L(x_0'(t)) \equiv 0$, which shows that $x_0'(t)$ satisfies (1.11). (Of course $x_0'(t)$ satisfies (1.11) even when $x_0(t)$ is constant, but it is then the trivial solution.)

Because (1.11) is so much simpler when $x_0(t)$ is constant, the entire study of this case is simpler. As a result we shall separate the cases of $x_0(t)$ constant and $x_0(t)$ variable, treating the former in Part II and the latter in Part III.

Let $\phi(t)$ and $\psi(t)$ be those solutions of (1.11) which satisfy the initial conditions $\phi(0) = 1$, $\phi'(0) = 0$, $\psi(0) = 0$, $\psi'(0) = 1$. The eigenvalues of the matrix

$$\begin{Vmatrix} \phi(L) & \psi(L) \\ \phi'(L) & \psi'(L) \end{Vmatrix} \tag{1.12}$$

are called the characteristic multipliers of (1.11). When $x_0(t)$ is constant, we have $L = T$, and the characteristic multipliers turn out to be $e^{r_1 T}$ and $e^{r_2 T}$ where r_1 and r_2 are the roots of the characteristic equation

$$r^2 + g_{x'}(x_0, 0)r + g_x(x_0, 0) = 0.$$

Except in the cases noted earlier, $e^{r_1 T}$ and $e^{r_2 T}$ are both different from $+1$.

When $x_0(t)$ is nonconstant, the characteristic multipliers may still be computed explicitly. The determinant of the matrix (1.12), which is the product of the characteristic multipliers, is the Wronskian of $\phi(t)$ and $\psi(t)$ evaluated at $t = L$, and so has the value

$$e^{-\int_0^L g_{x'}(x_0(t), x_0'(t))\,dt} \tag{1.13}$$

However, because $x_0'(t)$, a linear combination of $\phi(t)$ and $\psi(t)$ is periodic, one of the characteristic multipliers is equal to $+1$, so that the second has the value (1.13) which is in general different from $+1$. However, in the important case that $g(x, x')$ is independent of x', both characteristic multipliers are equal to $+1$.

We shall distinguish four cases, which we speak of as Cases I, II, III, and IV:

Case I: Neither characteristic multiplier of (1.11) is equal to $+1$.

Case II: Exactly one characteristic multiplier of (1.11) is equal to $+1$.

Case III: Both characteristic multipliers of (1.11) are equal to $+1$, but (1.12) is not the identity.

Case IV: Both characteristic multipliers of (1.11) are equal to $+1$, and (1.12) is the identity.

Another distinction between Cases III and IV is that in Case IV, all solutions of (1.11) are periodic with period L, while in Case III only $x_0'(t)$ and its constant multiples are periodic.

The cases are in increasing order of complexity. Case I can occur only when $x_0(t)$ is constant. A typical equation producing Case II is van der Pol's equation. A typical equation producing Case III is Duffing's equation. Case IV arises in the nearly linear equation with $g(x, x') \equiv k^2 x$.

1.3. The Method of Study. In all the cases studied, the general method of attack is the same. We assume that in (1.2) $g(x, x') \in C^2$ and $f(t, x, x', \varepsilon) \in C^1$. Let $x_0(t)$ be a periodic solution of (1.1), and choose the time origin so that $x_0'(0) = 0$. Let $x_0(0) = A$, and let the least period of $x_0(t)$ be L_0 if $x_0(t)$ is nonconstant. Let $f(t, x, x', \varepsilon)$ have least period T in t. If $x_0(t)$ is nonconstant, we assume L_0 is a rational multiple of T. Let L be the least common multiple of T and L_0 if $x_0(t)$ is variable; if $x_0(t)$ is constant, let $L = T$.

Let $x(t, \xi, \eta, \varepsilon)$ be the solution of (1.2) such that at $t = 0$, $x = A + \xi$, $x' = \eta$. This solution is unique because of the assumptions made about $g(x, x')$ and $f(t, x, x', \varepsilon)$. Also, $x(t, 0, 0, 0) \equiv x_0(t)$. Define $F(\xi, \eta, \varepsilon)$ and $G(\xi, \eta, \varepsilon)$ by

$$F(\xi, \eta, \varepsilon) = x(L, \xi, \eta, \varepsilon) - A - \xi,$$

$$G(\xi, \eta, \varepsilon) = x'(L, \xi, \eta, \varepsilon) - \eta.$$

Then $F(0, 0, 0) = G(0, 0, 0) = 0$, and $x(t, \xi, \eta, \varepsilon)$ has period L if and only if $F(\xi, \eta, \varepsilon) = G(\xi, \eta, \varepsilon) = 0$.

In each case the existence proof will consist of showing that for small values of ε the system

$$F(\xi, \eta, \varepsilon) = 0, \quad G(\xi, \eta, \varepsilon) = 0 \tag{1.14}$$

can be solved for ξ and η as functions of ε, with $\xi = \eta = 0$ when $\varepsilon = 0$, and the derivatives $d\xi/d\varepsilon$ and $d\eta/d\varepsilon$ finite at $\varepsilon = 0$. We give the explicit values of these derivatives where it is feasible, and otherwise show how to compute them explicitly. Knowing the values of $d\xi/d\varepsilon$ and $d\eta/d\varepsilon$ at $\varepsilon = 0$, we can determine the function $x_1(t)$ of (1.3) itself.

It will be possible in some cases to construct several terms of the perturbation series

$$x = x_0(t) + \varepsilon x_1(t) + \varepsilon^2 x_2(t) + \dots \tag{1.15}$$

This is done by substituting (1.15) formally into (1.2) and equating like powers of ε. The various functions $x_1(t), x_2(t), \ldots$ are found to satisfy linear equations with the same left member as (1.11), and the existence proof will give the possibility of finding $x_1(t), x_2(t), \ldots$ as periodic functions with period L. It would be quite difficult in general to start with the series (1.15), and to conclude existence of the periodic solution of (1.2) from the possibility of constructing (1.15). The series might not exist completely, and if it did, it might not be convergent, but merely asymptotic.

In the analysis of the system (1.14) it will be necessary to compute the values of the partial derivatives of F and G at (0, 0, 0). In Case I, and only in Case I, the Jacobian of F and G with respect to ξ and η at (0, 0, 0) is not zero. In this case we require only the first derivatives of F and G. In all other cases, this Jacobian is zero, and higher derivatives of F and G are required.

In Cases II, III, and IV, the best that can happen is that $d\xi/d\varepsilon$ and $d\eta/d\varepsilon$ at $\varepsilon = 0$ can be found from a knowledge of the second derivatives of F and G at (0, 0, 0). We shall go through this case completely. It will be required that a certain discriminant be positive, and this requirement will be equivalent to the non-vanishing of the critical Jacobian in the theory of Coddington and Levinson [3], Chapter 14, or of Friedrichs [5]. We shall study the case that third derivatives of F and G are required only when $x_0(t)$ is variable.

The determination of the partial derivatives of F and G at (0, 0, 0) is in general a long and tedious computation. We shall usually omit the computations. In a few cases the computation will be given to illustrate the technique. The partial derivatives of F and G are found from the corresponding partial derivatives of $x(t, \xi, \eta, \varepsilon)$ evaluated at $t = L$, $\xi = \eta = \varepsilon = 0$. The partial derivatives of x for $\xi = \eta = \varepsilon = 0$ satisfy linear differential equations having as common left member the left member of (1.11). These are found by differentiating (1.2) formally and then setting $\xi = \eta = \varepsilon = 0$.

In the following $g(x, x')$ and its partial derivatives are understood to be evaluated for $x = x_0(t)$, $x' = x_0'(t)$. $f(t, x, x', \varepsilon)$ and its partial derivatives are understood to be evaluated for $x = x_0(t)$, $x' = x_0'(t)$, $\varepsilon = 0$. The partial derivatives of $x(t, \xi, \eta, \varepsilon)$ are all understood to be evaluated for $\xi = \eta = \varepsilon = 0$. L( ) stands for the operator in the left member of equation (1.11).

With the above agreements, the first derivatives of $x(t, \xi, \eta, \varepsilon)$ at $\xi = \eta = \varepsilon = 0$ satisfy:

(1.16a) $$L(x_\xi) = 0, \quad x_\xi(0) = 1, \quad x_\xi'(0) = 0;$$

(1.16b) $$L(x_\eta) = 0, \quad x_\eta(0) = 0, \quad x_\eta'(0) = 1;$$

(1.16c) $$L(x_\varepsilon) = f, \quad x_\varepsilon(0) = x_\varepsilon'(0) = 0.$$

The second derivatives of $x(t, \xi, \eta, \varepsilon)$ at $\xi = \eta = \varepsilon = 0$ satisfy

(1.17a) $$L(x_{\xi\xi}) = -g_{xx}x_\xi^2 - 2g_{xx'}x_\xi x_\xi' - g_{x'x'}x_\xi^{2\prime};$$

(1.17b) $$L(x_{\xi\eta}) = -g_{xx}x_\xi x_\eta - g_{xx'}(x_\xi x_\eta' + x_\xi' x_\eta) - g_{x'x'}x_\xi' x_\eta';$$

(1.17c) $$L(x_{\eta\eta}) = -g_{xx}x_\eta^2 - 2g_{xx'}x_\eta x_\eta' - g_{x'x'}x_\eta^{2\prime};$$

(1.17d) $$L(x_{\xi\varepsilon}) = -g_{xx}x_\xi x_\varepsilon - g_{xx'}(x_\xi x_\varepsilon' + x_\xi' x_\varepsilon) - g_{x'x'}x_\xi' x_\varepsilon' + f_x x_\xi + f_{x'}x_\xi';$$

(1.17e) $$L(x_{\eta\varepsilon}) = -g_{xx}x_\eta x_\varepsilon - g_{xx'}(x_\eta x_\varepsilon' + x_\eta' x_\varepsilon) - g_{x'x'}x_\eta' x_\varepsilon' + f_x x_\eta + f_{x'}x_\eta';$$

(1.17f) $$L(x_{\varepsilon\varepsilon}) = -g_{xx}x_\varepsilon^2 - 2g_{xx'}x_\varepsilon x_\varepsilon' - g_{x'x'}x_\varepsilon^{2\prime} + 2f_x x_\varepsilon + 2f_{x'}x_\varepsilon' + 2f_\varepsilon.$$

Each second derivative of x satisfies the initial conditions $y(0) = y'(0) = 0$. The ten different third partial derivatives of x satisfy similar linear equations, which will not be written out.

On comparison of equations (1.16a) and (1.16b) with the definitions of $\phi(t)$ and $\psi(t)$, we see that

$$x_\xi(t, 0, 0, 0) \equiv \phi(t) \quad \text{and} \quad x_\eta(t, 0, 0, 0) \equiv \psi(t).$$

The first partial derivatives of F and G at $(0, 0, 0)$ are given by

$$
\begin{aligned}
F_\xi &= x_\xi(L) - 1 = \phi(L) - 1, \\
G_\xi &= x_\xi'(L) = \phi'(L), \\
F_\eta &= x_\eta(L) = \psi(L), \\
G_\eta &= x_\eta'(L) - 1 = \psi'(L) - 1, \\
F_\varepsilon &= x_\varepsilon(L), \\
G_\varepsilon &= x_\varepsilon'(L).
\end{aligned} \tag{1.18}
$$

If $x_{\alpha\beta}(t, 0, 0, 0)$ is any one of the six second partial derivatives of x, the corresponding second partial derivatives of F and G at $(0, 0, 0)$ are given by

$$
F_{\alpha\beta} = x_{\alpha\beta}(L), \quad G_{\alpha\beta} = x_{\alpha\beta}'(L). \tag{1.19}
$$

Higher derivatives of F and G are found similarly.

The Jacobian matrix of F and G with respect to ξ and η at $(0, 0, 0)$ is given by

$$
\left\| \begin{matrix} F_\xi & F_\eta \\ G_\xi & G_\eta \end{matrix} \right\| = \left\| \begin{matrix} \phi(L) - 1 & \psi(L) \\ \phi'(L) & \psi'(L) - 1 \end{matrix} \right\| \tag{1.20}
$$

In Case I, the rank of (1.20) is 2, in Cases II and III, the rank of (1.20) is 1, and in Case IV, the rank of (1.20) is 0.

1.4. Stability. We shall also consider the stability properties of the various periodic solutions that we find. Stability of a periodic solution in the form (1.3) will be described by the characteristic multipliers of the variation equation

$$
y'' + g_{x'}(x, x)y' + g_x(x, x')y = \varepsilon f_{x'}(t, x, x', \varepsilon)y' + \varepsilon f_x(t, x, x', \varepsilon)y. \tag{1.21}
$$

Characteristic multipliers of (1.21) are defined in terms of its solutions $\phi(t)$ and $\psi(t)$ just as for equation (1.11) (to which (1.21) reduces for $\varepsilon = 0$).

If both characteristic multipliers of (1.21) are less than 1 in absolute value, the periodic solution (1.3) will be called *completely stable.* It is at the same time *asymptotically stable,* which means that a solution of (1.2) with initial conditions near to those of the periodic solution will converge to the periodic solution as $t \to +\infty$. If both characteristic multipliers of (1.21) are greater than 1 in absolute value, the periodic solution will be called ***completely unstable.*** If both characteristic multipliers of (1.21) are of absolute value 1, but are not real, we will call the periodic solution *quasi-stable.* In this case, the stability is not completely determined by the variation equation. When the two characteristic multipliers have different absolute values, they are both real, since the entries in the matrix (1.12) are real. If they are both positive and are separated by +1, we call the periodic solution ***directly unstable,*** while if they are negative and separated by -1, the periodic solution is called ***inversely unstable.*** If either characteristic multiplier has the value +1 or -1, we do not classify stability. The various terminologies above are taken largely from Levinson's paper [7].

To study the characteristic multipliers of (1.21) we use strongly the fact that as $\varepsilon \to 0$, (1.21) reduces to (1.11), so that for small ε, the characteristic multipliers of (1.21) are near to those of (1.11). Thus in Case I, if neither characteristic multiplier of (1.11) has absolute value 1, the same will be true for small ε of the characteristic multipliers of (1.21), and the solution (1.3) will have the same stability properties (complete stability, complete instability, direct instability, inverse instability) as $x_0(t)$ has as a solution of (1.1). In Case II, where one characteristic multiplier of (1.11) is equal to +1, the corresponding characteristic multiplier of (1.21) will be real and near to +1, but usually different from +1 for $\varepsilon \neq 0$. Then from a knowledge of the second characteristic multiplier of (1.11) and a knowledge of the behavior of the characteristic multiplier near to +1, we can determine the stability of the periodic solution (1.3). It will be completely stable, completely unstable, or directly unstable. (The second characteristic multiplier of (1.11) and of (1.21) will have to be real, positive, and different from 1.) In Cases III and IV, when both characteristic multipliers of (1.11) are equal to +1, similar remarks apply as in Case II except that quasi-stability is also possible.

In Cases II, III, and IV, where it is necessary to determine the way in which the characteristic multipliers near +1 behave, we rewrite (1.21) as

$$
\begin{aligned}
L(y) &= \varepsilon(-g_{xx}x_1 y - g_{xx'}x_1 y' - g_{xx'}x_1'y - g_{x'x'}x_1'y' \\
&\quad + f_x y + f_{x'}y') + o(\varepsilon)
\end{aligned} \tag{1.22}
$$

where in (1.22) $L(\ )$ has the same meaning as in (1.11), the derivatives of g are evaluated for $x = x_0(t)$, $x' = x_0'(t)$, and the derivatives of f are evaluated for $x = x_0(t)$, $x' = x_0'(t)$, and $\varepsilon = 0$. Denote the fundamental solutions of (1.11) by $\phi_0(t)$ and $\psi_0(t)$. Then using $\phi_0(t)$ and $\psi_0(t)$, (1.22) can be transformed to an integral equation. Using $\phi_0(t)$ and $\psi_0(t)$ as initial approximations, the solutions $\phi(t)$ and $\psi(t)$ of (1.21) or (1.22) can be determined for small ε, so that the values of $\phi(L)$, $\phi'(L)$, $\psi(L)$ and $\psi'(L)$ can be determined for small ε. Thus the characteristic multipliers of (1.21) can be studied for small ε. In each case, the same partial derivatives of F and G which are required to establish existence of the periodic solution (1.3) are sufficient to determine its stability for small ε.

1.5. Autonomous Perturbations. The case that in equation (1.2), $f(t, x, x', \varepsilon)$ is in fact independent of t introduces extra difficulties. When $x_0(t)$ is nonconstant, there are still periodic solutions of the perturbed system near to $x_0(t)$, but they will have, as a rule, periods which differ from that of $x_0(t)$. This difficulty does not arise when $x_0(t)$ is constant, since in that case the corresponding periodic solutions of the perturbed system will also be constant.

The method of this paper will apply to the case of autonomous perturbations. The case of autonomous perturbations with $x_0(t)$ constant is studied in Part II, while the case with $x_0(t)$ nonconstant is treated in Part IV.

PART II: PERTURBATION OF A CONSTANT SOLUTION

2.1. Introduction. In this part of the paper we assume the periodic solution $x_0(t)$ of (1.1) to be identically constant. Without loss of generality we may assume that $x_0(t) \equiv 0$. This implies that $g(0,0) = 0$. We seek periodic solutions of (1.2) of period T having the form

$$x = \varepsilon x_1(t) + o(\varepsilon). \tag{2.1}$$

The associated variational equation is

$$L(y) \equiv y' + g_{x'}(0,0)y' + g_x(0,0)y = 0. \tag{2.2}$$

The two characteristic multipliers of (2.2) are $e^{r_1 T}$ and $e^{r_2 T}$, where r_1 and r_2 are the roots of

$$r^2 + g_{x'}(0,0)r + g_x(0,0) = 0.$$

It follows that we have Case I, where neither characteristic multiplier is equal to $+1$, when $g_{x'}(0,0) \neq 0$ and $g_x(0,0) \neq 0$, or when $g_{x'}(0,0) = 0$ and $g_x(0,0) \neq 4n^2\pi^2 T^{-2}$ for all integers $n = 0, \pm 1, \pm 2, \ldots$. We have Case II, where just one of the characteristic multipliers is equal to $+1$, when $g_{x'}(0,0) \neq 0$ and $g_x(0,0) = 0$. We have Case III, where both characteristic multipliers are equal to $+1$, but not all solutions of (2.2) have period T, when $g_{x'}(0,0) = 0$, and $g_x(0,0) = 0$. We have Case IV, where all solutions of (2.2) have period T, when $g_{x'}(0,0) = 0$ and $g_x(0,0) = 4n^2\pi^2 T^{-2}$ for some nonzero integer n.

On substituting (2.1) into (1.2) and equating coefficients of like powers of ε, we find that $x_1(t)$ must be a periodic solution with period T of

$$y'' + g_{x'}(0,0)y' + g_x(0,0)y = f(t,0,0,0). \tag{2.3}$$

In Case I such a solution exists uniquely. In Cases II, III, and IV, certain orthogonality conditions must be imposed on $f(t,0,0,0)$.

Note: Throughout Part II we shall give the values of the various partial derivatives of F and G at $(0,0,0)$ without giving computations, except for a few sample computations in Case III.

2.2. Existence and Stability in Case I.

THEOREM 2.1. *In the real differential equation*

$$x'' + g(x, x') = \varepsilon f(t, x, x', \varepsilon) \tag{2.4}$$

let $g(x, x') \in C^1$ *and* $f(t, x, x', \varepsilon) \in C$. *Let* $x \equiv 0$ *be a solution of* (2.4) *when* $\varepsilon = 0$. *Let* $f(t, x, x', \varepsilon)$ *have least period* T *in* t. *Let the variation equation*

$$y'' + g_{x'}(0, 0)y' + g_x(0, 0)y = 0 \tag{2.5}$$

have no nontrivial solution of period T.

Then for small nonzero ε (2.4) *has a unique solution of period* T *of the form*

$$x = \varepsilon x_1(t) + o(\varepsilon) \tag{2.6}$$

where $x_1(t)$ *is the unique solution of*

$$y'' + g_{x'}(0, 0)y' + g_x(0, 0)y = f(t, 0, 0, 0) \tag{2.7}$$

of period T.

If neither characteristic multiplier of (2.5) *has absolute value* 1, *then for small nonzero* ε, *neither characteristic multiplier of*

$$y'' + g_{x'}(x, x')y' + g_x(x, x')y = \varepsilon f_{x'}(t, x, x', \varepsilon)y' + \varepsilon f_x(t, x, x', \varepsilon)y \tag{2.8}$$

has absolute value 1, *and the stability of* x *is the same as that of* $x \equiv 0$ *when* $\varepsilon = 0$. *In fact* x *is completely stable if* $g_x(0, 0) > 0$ *and* $g_x(0, 0) > 0$, *completely unstable if* $g_{x'}(0, 0) < 0$ *and* $g_x(0, 0) > 0$, *and directly unstable if* $g_x(0, 0) < 0$.

Proof of Theorem 2.1: The existence and uniqueness of the periodic solution follow directly from Theorem 1.1, Chapter 14 of [3], but we shall give a proof using the technique indicated in Part I.

Let $\phi(t)$ and $\psi(t)$ be those solutions of (2.5) with $\phi(0) = \psi'(t) = 1$, $\phi'(0) = \psi(0) = 0$. Let $W(t) \equiv \phi(t)\psi'(t) - \psi(t)\phi'(t)$ be their Wronskian. ($W(t) \equiv \exp[-g_{x'}(0, 0)t]$.) Then at $(0, 0, 0)$ the first partial derivatives of F and G turn out to have the values

$$F_\xi = \phi(T) - 1, \quad F_\eta = \psi(T)$$

$$G_\xi = \phi'(T), \quad G_\eta = \psi'(T) - 1,$$

(2.9)

$$F_\varepsilon = \int_0^T \frac{\phi(s)\psi(T) - \psi(s)\phi(T)}{W(s)} f(s, 0, 0, 0)\, ds,$$

$$G_\varepsilon = \int_0^T \frac{\phi(s)\psi'(T) - \psi(s)\phi'(T)}{W(s)} f(s, 0, 0, 0)\, ds.$$

By the Case I hypothesis, the Jacobian of F and G with respect to ξ and η is not zero. Hence, by the implicit function theorem, we can find ξ and η uniquely as functions of ε for small ε. This proves the existence and uniqueness of the periodic solution.

As was remarked in Section 2.1, $x_1(t)$ must be a solution of equation (2.7) with period T. But since the homogeneous equation (2.5) has no nontrivial solution of period T, (2.7) has a unique periodic solution of period T, which must be $x_1(t)$.

If neither characteristic multiplier of (2.5) has absolute value 1, then by continuity neither characteristic multiplier of (2.8) has absolute value 1 for small ε, and the absolute values of the characteristic multipliers of (2.8) bear the same relation to 1 as do those of (2.5). The solution $x \equiv 0$ for $\varepsilon = 0$ is clearly completely stable if r_1 and r_2 both have negative real part, and completely unstable if r_1 and r_2 both have positive real part, and directly unstable if r_1 and r_2 are of opposite sign. Hence the stability is as asserted if $g_{x'}(0, 0)$ and $g_x(0, 0)$ have the indicated signs. This completes the proof of Theorem 2.1.

Remarks: The investigation of stability is more difficult in the case omitted, that when $g_{x'}(0, 0) = 0$, and $g_x(0, 0) \neq 4n^2\pi^2T^{-2}$, when both characteristic multipliers of (2.5) have absolute value 1. This will be investigated in Section 2.7 using techniques developed for Cases II, III, and IV.

Since for the periodic solution $\xi = x(0) = \varepsilon x_1(0) + o(\varepsilon)$, and $\eta = x'(0) = \varepsilon x_1'(0) + o(\varepsilon)$, we have that the derivatives $d\xi/d\varepsilon$ and $d\eta/d\varepsilon$ at $\varepsilon = 0$ satisfy

$$\frac{d\xi}{d\varepsilon} = x_1(0), \quad \frac{d\eta}{d\varepsilon} = x_1'(0).$$

On the other hand, these derivatives can be computed from (2.9) to obtain

$$\frac{d\xi}{d\varepsilon} = -\frac{F_\varepsilon G_\eta - F_\eta G_\varepsilon}{F_\xi G_\eta - F_\eta G_\xi}, \quad \frac{d\eta}{d\varepsilon} = -\frac{F_\xi G_\varepsilon - F_\varepsilon G_\xi}{F_\xi G_\eta - F_\eta G_\xi}.$$

It is readily verified that these latter expressions are indeed the initial values of the unique periodic solution of (2.7).

2.3. Existence and Stability in Case II. We saw in Section 2.1 that Case II occurs if and only if $g_{x'}(0,0) \neq 0$ and $g_x(0,0) = 0$. We shall denote the value of $g_{x'}(0,0)$ by α. The variation equation then takes the simple form

$$y'' + \alpha y' = 0. \tag{2.10}$$

The characteristic multipliers of (2.10) are 1 and $e^{-\alpha T}$.

In the discussion which follows, it will be convenient to introduce abbreviations. The second partial derivatives of $g(x, x')$ at $(0, 0)$ will be written without the designation $(0, 0)$. Thus $g_{xx}(0, 0)$ will be written simply g_{xx}. The function $f(t, x, x', \varepsilon)$ and its partial derivatives when evaluated for $x = x' = \varepsilon = 0$ will be written without showing the three zeros. Thus $f_x(t, 0, 0, 0)$ will be written simply $f_x(t)$. We shall continue to write the partial derivatives of F and G at $(0, 0, 0)$ without writing out the zeros.

The solutions $\phi(t)$ and $\psi(t)$ of (2.10) are

$$\phi(t) = 1, \quad \psi(t) = \frac{1 - e^{-\alpha t}}{\alpha}.$$

The first partial derivatives of F and G at $(0, 0, 0)$ have the values

$$F_\xi = 0, \quad F_\eta = \frac{1 - e^{-\alpha T}}{\alpha}, \quad F_\varepsilon = \frac{1}{\alpha}\int_0^T f(t)\,dt - \frac{1}{\alpha}\int_0^T e^{-\alpha(T-s)} f(s)\,ds,$$

$$G_\xi = 0, \quad G_\eta = e^{-\alpha T} - 1, \quad G_\varepsilon = \int_0^T e^{-\alpha(T-s)} f(s)\,ds.$$

The Jacobian of F and G with respect to ξ and η is zero. The rank of the Jacobian matrix is 1, since neither F_η nor G_η vanishes. In order that $d\xi/d\varepsilon$ and $d\eta/d\varepsilon$ exist, it is necessary that the rank of the matrix

$$\left\| \begin{matrix} F_\xi & F_\eta & F_\varepsilon \\ G_\xi & G_\eta & G_\varepsilon \end{matrix} \right\|$$

also be 1. But this requirement is equivalent to

$$\int_0^T f(t, 0, 0, 0)\, dt = 0. \tag{2.11}$$

If we make this requirement, the table of first derivatives becomes

$$\begin{aligned} &F_\xi = 0, \quad F_\eta = \frac{1 - e^{-\alpha T}}{\alpha}, \quad F_\varepsilon = -\frac{1}{\alpha}\int_0^T e^{-\alpha(T-s)} f(s)\, ds, \\ &G_\xi = 0, \quad G_\eta = e^{-\alpha T} - 1, \quad G_\varepsilon = \int_0^T e^{-\alpha(T-s)} f(s)\, ds. \end{aligned} \tag{2.12}$$

From (2.12) we can obtain the value of $d\eta/d\varepsilon$ at $\varepsilon = 0$. It is the common value of $-F_\varepsilon/F_\eta$ and $-G_\varepsilon/G_\eta$ and is

$$\frac{d\eta}{d\varepsilon} = \frac{1}{1 - e^{-\alpha T}} \int_0^T e^{-\alpha(T-s)} f(s)\, ds = \frac{1}{e^{\alpha T} - 1} \int_0^T e^{\alpha s} f(s)\, ds \tag{2.13}$$

However, to compute the derivative $d\xi/d\varepsilon$ at $\varepsilon = 0$, it will be necessary to use the second partial derivatives of F and G at $(0, 0, 0)$.

To compute $d\xi/d\varepsilon$ we eliminate η between the equations $F(\xi, \eta, \varepsilon) = 0$ and $G(\xi, \eta, \varepsilon) = 0$. Since $F_\eta \neq 0$, we can solve the equation $F(\xi, \eta, \varepsilon) = 0$ for η as a function of η and ε near $\xi = \varepsilon = 0$. If the result is $\eta = H(\xi, \varepsilon)$, we then get our relation between ξ and ε alone by substituting $H(\xi, \varepsilon)$ for η in $G(\xi, \eta, \varepsilon) = 0$. If we set

$$J(\xi, \varepsilon) = G(\xi, H(\xi, \varepsilon), \varepsilon),$$

we have to solve the equation $J(\xi, \varepsilon) = 0$ for ξ as a function of ε near $\varepsilon = 0$.

The second partial derivatives of F and G at $(0, 0, 0)$ have the values:

$$F_{\xi\xi} = -\frac{g_{xx}}{\alpha^2}(e^{-\alpha T} + \alpha T - 1),$$

$$G_{\xi\xi} = \frac{g_{xx}}{\alpha}(e^{-\alpha T} - 1),$$

$$F_{\xi\eta} = -\frac{g_{xx}}{\alpha^3}(\alpha T e^{-\alpha T} + 2e^{-\alpha T} + \alpha T - 2) + \frac{g_{xx'}}{\alpha^2}(\alpha T e^{-\alpha T} + e^{-\alpha T} - 1),$$

$$G_{\xi\eta} = \frac{g_{xx}}{\alpha^2}(\alpha T e^{-\alpha T} + e^{-\alpha T} - 1) - \frac{g_{xx'}}{\alpha}(\alpha T e^{-\alpha T}),$$

$$F_{\eta\eta} = -\frac{g_{xx}}{\alpha^4}(\frac{1}{2}e^{-2\alpha T} + 2\alpha T e^{-\alpha T} + 2e^{-\alpha T} + \alpha T - \frac{5}{2}) + \frac{2g_{xx'}}{\alpha^3}(\frac{1}{2}e^{-2\alpha T}$$
$$+ \alpha T e^{-\alpha T} - \frac{1}{2}) - \frac{g_{x'x'}}{\alpha^2}(\frac{1}{2}e^{-2\alpha T} - e^{-\alpha T} + \frac{1}{2}),$$

$$G_{\eta\eta} = \frac{g_{xx}}{\alpha^3}(e^{-2\alpha T} + 2\alpha T e^{-\alpha T} - 1) - \frac{2g_{xx'}}{\alpha^2}(e^{-2\alpha T} + \alpha T e^{-\alpha T} - e^{-\alpha T})$$
$$+ \frac{g_{x'x'}}{\alpha}(e^{-2\alpha T} - e^{-\alpha T}),$$

(2.14)
$$F_{\xi\varepsilon} = -\frac{g_{xx}}{\alpha^3}\int_0^T [\alpha(T-s)e^{-\alpha(T-s)} + 2e^{-\alpha(T-s)} - \alpha s] f(s)\,ds$$
$$+ \frac{g_{xx'}}{\alpha^2}\int_0^T [\alpha(T-s)e^{-\alpha(T-s)} + e^{-\alpha(T-s)}] f(s)\,ds$$
$$- \frac{1}{\alpha}\int_0^T [e^{-\alpha(T-s)} - 1] f_x(s)\,ds,$$

$$G_{\xi\varepsilon} = \frac{g_{xx}}{\alpha^2}\int_0^T [\alpha(T-s)e^{-\alpha(T-s)} + e^{-\alpha(T-s)}] f(s)\,ds$$
$$- \frac{g_{xx'}}{\alpha}\int_0^T [\alpha(T-s)e^{-\alpha(T-s)}] f(s)\,ds + \int_0^T e^{-\alpha(T-s)} f_x(s)\,ds,$$

$$F_{\eta\varepsilon} = -\frac{g_{xx}}{\alpha^4}\int_0^T [\frac{1}{2}e^{-\alpha(2T-s)} + \alpha(T-s)e^{-\alpha(T-s)} - \alpha s e^{-\alpha T} + 2e^{-\alpha(T-s)}$$
$$- \frac{1}{2}e^{-\alpha s} - \alpha s] f(s)\,ds + \frac{g_{xx'}}{\alpha^3}\int_0^T [e^{-\alpha(2T-s)}$$
$$+ \alpha(T-s)e^{-\alpha(T-s)} - \alpha s e^{-\alpha T} + e^{-\alpha(T-s)}] f(s)\,ds$$
$$- \frac{g_{x'x'}}{\alpha^2}\int_0^T [\frac{1}{2}e^{-\alpha(2T-s)} + \frac{1}{2}e^{-\alpha s}] f(s)\,ds$$

$$+\frac{1}{\alpha^2}\int_0^T(1-e^{-\alpha s}-e^{-\alpha(T-s)}+e^{-\alpha T})f_x(s)\,ds$$

$$+\frac{1}{\alpha}\int_0^T(e^{-\alpha s}-e^{-\alpha T})f_{x'}(s)\,ds,$$

$$G_{\eta\varepsilon}=\frac{g_{xx}}{\alpha^3}\int_0^T[e^{-\alpha(2T-s)}+\alpha(T-s)e^{-\alpha(T-s)}+e^{-\alpha(T-s)}$$

$$-\alpha s e^{-\alpha T}]f(s)\,ds-\frac{g_{xx'}}{\alpha^2}\int_0^T[2e^{-\alpha(2T-s)}$$

$$+\alpha(T-s)e^{-\alpha(T-s)}-\alpha s e^{-\alpha T}]f(s)\,ds$$

$$+\frac{g_{x'x'}}{\alpha}\int_0^T e^{-\alpha(2T-s)}f(s)\,ds$$

$$+\frac{1}{\alpha}\int_0^T[e^{-\alpha(T-s)}-e^{-\alpha T}]f_x(s)\,ds+\int_0^T e^{-\alpha T}f_{x'}(s)\,ds,$$

$$F_{\varepsilon\varepsilon}=-\frac{2g_{xx}}{\alpha^4}\int_0^T\int_0^s[\tfrac{1}{2}e^{-\alpha(2T-s-u)}+\alpha(T-s)e^{-\alpha(T-s)}+\alpha(T-s)e^{-\alpha(T-u)}$$

(2.14) (continued)

$$-\frac{1}{2}e^{-\alpha(s-u)}+2e^{-\alpha(T-s)}+\alpha(T-s)-2]f(u)f(s)\,du\,ds$$

$$+\frac{2g_{xx'}}{\alpha^3}\int_0^T\int_0^s[e^{-\alpha(2T-s-u)}+\alpha(T-s)e^{-\alpha(T-s)}+\alpha(T-s)e^{-\alpha(T-u)}$$

$$+e^{-\alpha(T-s)}-e^{-\alpha(T-u)}]f(u)f(s)\,du\,ds$$

$$-\frac{2g_{x'x'}}{\alpha^2}\int_0^T\int_0^s[\tfrac{1}{2}e^{-\alpha(2T-s-u)}+\tfrac{1}{2}e^{-\alpha(s-u)}-e^{-\alpha(T-u)}]$$

$$f(u)f(s)\,du\,ds+\frac{2}{\alpha^2}\int_0^T\int_0^s[e^{-\alpha(T-u)}-e^{-\alpha(T-s)}$$

$$-e^{-\alpha(s-u)}+1]f_x(s)f(u)\,du\,ds+\frac{2}{\alpha}\int_0^T\int_0^s[e^{-\alpha(s-u)}$$

$$-e^{-\alpha(T-u)}]f_{x'}(s)f(u)\,du\,ds-\frac{2}{\alpha}\int_0^T[e^{-\alpha(T-s)}-1]f_\varepsilon(s)\,ds$$

$$G_{\varepsilon\varepsilon}=\frac{2g_{xx}}{\alpha^3}\int_0^T\int_0^s[e^{-\alpha(2T-s-u)}+\alpha(T-s)e^{-\alpha(T-s)}+\alpha(T-s)e^{-\alpha(T-u)}$$

$$+e^{-\alpha(T-s)}-e^{-\alpha(T-u)}-1]f(u)f(s)\,du\,ds$$

$$- \frac{2g_{xx'}}{\alpha^2} \int_0^T \int_0^s [2e^{-\alpha(2T-s-u)} + \alpha(T-s)\, e^{-\alpha(T-s)} + \alpha(T-s)\, e^{-\alpha(T-u)}$$

$$- 2e^{-\alpha(T-u)}]\, f(u) f(s)\, duds + \frac{2g_{x'x'}}{\alpha} \int_0^T \int_0^s [e^{-\alpha(2T-s-u)}$$

(2.14) (continued)

$$- e^{-\alpha(T-u)}]\, f(s) f(u)\, duds + \frac{2}{\alpha} \int_0^T \int_0^s [e^{-\alpha(T-s)} - e^{-\alpha(T-u)}]$$

$$f_x(s) f(u)\, duds + 2 \int_0^T \int_0^s e^{-\alpha(T-u)} f_{x'}(s) f(u)\, duds$$

$$+ 2 \int_0^T e^{-\alpha(T-s)} f_\varepsilon(s)\, ds.$$

Using the values in (2.12) and (2.14) we can compute the first and second derivatives of the function $J(\xi, \varepsilon)$ at $(0, 0)$. The results are:

$$J_\xi = 0, \quad J_\varepsilon = 0,$$

$$J_{\xi\xi} = - Tg_{xx},$$

$$J_{\xi\varepsilon} = \int_0^T f_x(s)\, ds + \frac{g_{xx}}{\alpha} \int_0^T sf(s)\, ds - \frac{g_{xx}}{\alpha} \frac{T}{e^{-\alpha T} - 1} \int_0^T e^{\alpha s} f(s)\, ds,$$

$$J_{\varepsilon\varepsilon} = \frac{g_{xx}}{\alpha^3} \left[\frac{-\alpha T[\int_0^T e^{\alpha s} f(s)\, ds]^2}{(e^{-\alpha T} - 1)^2} + \frac{\int_0^T e^{\alpha s} f(s)\, ds \int_0^T e^{-\alpha s} f(s)\, ds}{e^{\alpha T} - 1} \right.$$

(2.15)

$$+ \frac{2\alpha \int_0^T e^{\alpha s} f(s)\, ds \int_0^T sf(s)\, ds}{e^{\alpha T} - 1} + 2\alpha \int_0^T \int_0^s sf(s) f(u)\, duds$$

$$\left. + \int_0^T \int_0^s e^{-\alpha(s-u)} f(s) f(u)\, duds \right]$$

$$- \frac{g_{x'x'}}{\alpha} \left[\frac{\int_0^T e^{\alpha s} f(s)\, ds \int_0^T e^{-\alpha s} f(s)\, ds}{e^{\alpha T} - 1} \right.$$

$$\left. + \int_0^T \int_0^s e^{-\alpha(s-u)} f(s) f(u)\, duds \right]$$

$$+\frac{2}{\alpha}\frac{\int_0^T(1-e^{-\alpha s})f_x(s)\,ds\int_0^T e^{\alpha s}f(s)\,ds}{e^{\alpha T}-1}$$

(2.15) (continued)

$$+\frac{2}{\alpha}\int_0^T\int_0^s(1-e^{-\alpha(s-u)})f_x(s)f(u)\,du\,ds$$

$$+\frac{2\int_0^T e^{\alpha s}f(s)\,ds\int_0^T e^{-\alpha s}f_{x'}(s)\,ds}{e^{\alpha T}-1}$$

$$+2\int_0^T\int_0^s e^{-\alpha(s-u)}f_{x'}(s)f(u)\,du\,ds+2\int_0^T f_\varepsilon(s)\,ds.$$

In the computation of (2.14) and (2.15), use was frequently made of (2.11).

We now have that near $\xi = \varepsilon = 0$,

$$J(\xi,\varepsilon) = \frac{1}{2}J_{\xi\xi}\xi^2 + J_{\xi\varepsilon}\xi\varepsilon + \frac{1}{2}J_{\varepsilon\varepsilon}\varepsilon^2$$
$$+ \text{ higher order terms.}$$

If the discriminant

(2.16) $$J_{\xi\varepsilon}^2 - J_{\xi\xi}J_{\varepsilon\varepsilon}$$

is positive, the equation

(2.17) $$\frac{1}{2}J_{\xi\xi}m^2 + J_{\xi\varepsilon}m + \frac{1}{2}J_{\varepsilon\varepsilon} = 0$$

has two real simple roots if $J_{\xi\xi} \neq 0$, and one simple real root of $J_{\xi\xi} = 0$. Corresponding to each such simple real root, a modification of the implicit function theorem, (Cf. [9], Lemma 8, page 508) shows that there is a solution $\xi = \xi(\varepsilon)$ of the equation $J(\xi,\varepsilon) = 0$ with $\xi = 0$ and $d\xi/d\varepsilon = m$ at $\varepsilon = 0$.

Thus if the discriminant (2.16) is positive, we are able to solve the system $F(\xi,\eta,\varepsilon) = G(\xi,\eta,\varepsilon) = 0$ for ξ and η as functions of ε near $\varepsilon = 0$. Moreover we can find the values at $\varepsilon = 0$ of $d\xi/d\varepsilon$ and $d\eta/d\varepsilon$, the former from the roots of (2.17) and the latter from (2.13). This proves the existence of one or two periodic solutions of (1.2) in the form (1.3). The function $x_1(t)$ is completely determined from equation (2.3) and the fact that $x_1(0) = d\xi/d\varepsilon$ and $x_1'(0) = d\xi/d\varepsilon$, the derivatives being evaluated at $\varepsilon = 0$.

It is possible to determine $x_1(t)$ in a formal manner from the perturbation series. Suppose we set

$$x = \varepsilon x_1(t) + \varepsilon^2 x_2(t) + \ldots$$

and substitute into equation (1.2). When coefficients of like powers of ε are equated, a series of linear equations for the various functions $x_1(t), x_2(t)$, are found. The equation for $x_1(t)$ was seen earlier to be (2.3). The equation for $x_2(t)$ is

$$x_2'' + g_{x'}x_2' + g_x x_2 = -\frac{1}{2}g_{xx}x_1^2 - g_{xx'}x_1x_1' - \frac{1}{2}g_{x'x'}x_1'^2 \tag{2.18}$$
$$+ f_x x_1 + f_{x'}x_1' + f_\varepsilon .$$

The derivatives of g and f in (2.18) are all evaluated at $x = x' = \varepsilon = 0$.

Now $x_1(t)$ must have period T, and because of (2.11), equation (2.3) does have solutions of period T. They are all given by

$$y = C_1 + \frac{(1 - e^{-\alpha t})\int_0^T e^{\alpha s} f(s)\,ds}{\alpha(e^{\alpha T} - 1)} + \frac{1}{\alpha}\int_0^t (1 - e^{-\alpha(t-s)}) f(s)\,ds \tag{2.19}$$
$$= C_1 + h_1(t).$$

In (2.19), C_1 is an arbitrary constant. The function $h_1(t)$ is the unique periodic solution of (2.3) having $y(0) = 0$. (Of course (2.3) is here adapted to our Case II notation.) We have thus found that $x_1(t)$ must have the form

$$x_1(t) = C_1 + h_1(t), \tag{2.20}$$

where C_1 is as yet undetermined. It can be verified that $x_1'(0) = h_1'(0)$ has the value req uired by (2.13).

Now in order that (2.18) have a solution of period T, its right member must have mean value zero. Writing this out as an integral, and using (2.20) for x_1, we find a quadratic equation for C_1, and in some cases a linear equation. It turns out that the equation for C_1 is equivalent to equation (2.17), so that we may write

$$\frac{1}{2}J_{\xi\xi}C_1^2 + J_{\xi\varepsilon}C_1 + \frac{1}{2}J_{\varepsilon\varepsilon} = 0. \tag{2.21}$$

This shows that C_1 will have the same value or values as $d\xi/d\varepsilon$, and that under the hypotheses we shall make, C_1 will be a simple real root of (2.21).

THEOREM 2.2. *In equation* (1.2) *let* $g(x, x') \in C^2$ *and* $f(t, x, x', \varepsilon) \in C^1$. *Let equation* (1.1) *have the constant solution* $x \equiv 0$. *In addition assume that*

$$g_{x'}(0, 0) = \alpha \neq 0, \tag{2.22}$$

$$g_x(0, 0) = 0, \tag{2.23}$$

$$\int_0^T f(t, 0, 0, 0)\, dt = 0 \tag{2.24}$$

$$J_{\xi\varepsilon}^2 - J_{\xi\xi} J_{\varepsilon\varepsilon} > 0. \tag{2.25}$$

Then equation (1.2) *has one or two periodic solutions of the form*

$$x = \varepsilon x_1(t) + o(\varepsilon) \tag{2.26}$$

according as $g_{xx}(0, 0) = 0$ *or* $g_{xx}(0, 0) \neq 0$. *In each case the function* $x_1(t)$ *is a uniquely determined periodic solution of*

$$y'' + \alpha y' = f(t, 0, 0, 0). \tag{2.27}$$

Proof of Theorem 2.2: The existence has already been proved in the discussion preceding the theorem. By (2.15) the condition that $g_{xx}(0, 0) = 0$ is equivalent to $J_{\xi\xi} = 0$. We have seen that $x_1(t)$ must have the form

$$x_1(t) = C_1 + h_1(t),$$

and that C_1 is determined by equation (2.21). Thus C_1 is equal to $d\xi/d\varepsilon$ at $\varepsilon = 0$. This completes the proof.

Remark: If $g_{xx}(0, 0) = 0$, and therefore in the case that $g(x, x')$ is linear, the conditions simplify a great deal. In particular, condition (2.25) reduces to

$$\int_0^T f_x(t, 0, 0, 0)\, dt \neq 0.$$

There is only one periodic solution in this case.

We now determine the stability of the periodic solutions found in Theorem 2.2. To do this, we study the related variational equation

$$y'' + g_{x'}(x, x')y' + g_x(x, x')y = \varepsilon f_{x'}(t, x, x', \varepsilon)y' + \varepsilon f_x(t, x, x', \varepsilon)y. \tag{2.28}$$

Since equation (2.28) reduces to (2.10) for $\varepsilon = 0$, one of the characteristic multipliers of (2.28) will be near $e^{-\alpha T}$ for small ε, so that its absolute value will not be equal to 1 for small ε. The second characteristic multiplier of (2.28) will be equal to 1 for $\varepsilon = 0$, but will be different from 1 for nonzero ε. To determine the stability of the periodic solution $x(t)$, we shall have to determine the behavior for small nonzero ε of this second characteristic multiplier. We shall evaluate the derivative with respect to ε of this characteristic multiplier for $\varepsilon = 0$.

If we set

$$x = \varepsilon x_1(t) + o(\varepsilon),$$

and substitute in (2.28), we obtain:

$$\begin{aligned} y'' + \alpha y' &= \varepsilon\{[-g_{xx}x_1(t) - g_{xx'}x_1'(t) + f_x(t)]y \\ &\quad + [-g_{xx'}x_1(t) - g_{x'x'}x_1'(t) + f_{x'}(t)]y'\} + o(\varepsilon). \end{aligned}$$

In (2.29) the derivatives of g and f are evaluated for $x = x' = \varepsilon = 0$.

Let $\phi(t)$ be that solution of (2.28) which has $\phi(0) = 1$ and $\phi'(0) = 0$. Then $\phi(t)$ satisfies the integral equation

$$\begin{aligned} \phi(t) &= 1 + \varepsilon\int_0^t \frac{1-e^{-\alpha(t-s)}}{\alpha}\{[-g_{xx}x_1(s) - g_{xx'}x_1'(s) + f_x(s)]\phi(s) \\ &\quad + [-g_{xx'}x_1(s) - g_{x'x'}x_1'(s) + f_{x'}(s)]\phi'(s)\}ds + o(\varepsilon). \end{aligned} \tag{2.30}$$

Using $\phi = 1$ as an initial approximation, this gives

$$\phi(t) = 1 + \varepsilon\int_0^t \frac{1-e^{-\alpha(t-s)}}{\alpha}[-g_{xx}x_1'(s) + f_x(s)]ds + o(\varepsilon). \tag{2.31}$$

Differentiating we have

$$\phi'(t) = \varepsilon\int_0^t e^{-\alpha(t-s)}[-g_{xx}x_1(s) - g_{xx'}x_1'(s) + f_x(s)]ds + o(\varepsilon). \tag{2.32}$$

Setting $t = T$ we obtain

$$\begin{aligned} \phi(T) &= 1 + \varepsilon\int_0^T \frac{1-e^{-\alpha(T-s)}}{\alpha}[-g_{xx}x_1(s) - g_{xx'}x_1'(s) \\ &\quad + f_x(s)]ds + o(\varepsilon) = 1 + \varepsilon d_{11} + o(\varepsilon), \end{aligned} \tag{2.33}$$

$$\phi'(T) = \varepsilon \int_0^T e^{-\alpha(T-s)} [-g_{xx} x_1(s) - g_{xx'} x_1'(s) + f_x(s)]\,ds \tag{2.34}$$
$$+ o(\varepsilon) = \varepsilon d_{21} + o(\varepsilon).$$

In a similar manner, if $\psi(t)$ is that solution of (2.28) with $\psi(0) = 0$ and $\psi'(0) = 1$, we find

$$\psi(T) = \frac{1 - e^{-\alpha T}}{\alpha} + \varepsilon d_{12} + o(\varepsilon), \tag{2.35}$$

$$\psi'(T) = e^{-\alpha T} + \varepsilon d_{22} + o(\varepsilon). \tag{2.36}$$

The explicit values of d_{12} and d_{22} are not needed. The equation whose roots are the characteristic multipliers of (2.28) is

$$G(\varepsilon, \lambda) = 0 \tag{2.37}$$

where

$$G(\varepsilon, \lambda) = \det \begin{Vmatrix} \phi(T) - \lambda & \psi(T) \\ \phi'(T) & \psi'(T) - \lambda \end{Vmatrix}$$
$$= \det \begin{Vmatrix} 1 - \lambda + \varepsilon d_{11} + o(\varepsilon) & \dfrac{1-e^{-\alpha T}}{\alpha} + \varepsilon d_{12} + o(\varepsilon) \\ \varepsilon d_{21} + o(\varepsilon) & e^{-\alpha T} - \lambda + \varepsilon d_{22} + o(\varepsilon) \end{Vmatrix} \tag{2.38}$$

Since $\lambda = 1$ is a characteristic multiplier when $\varepsilon = 0$, $G(0, 1) = 0$. If $\lambda(\varepsilon)$ denotes that root of (2.37) which is 1 for $\varepsilon = 0$, we can obtain the derivative $\lambda'(0)$ from the relation

$$\lambda'(0) = -\frac{G_\varepsilon(0, 1)}{G_\lambda(0, 1)}. \tag{2.39}$$

But the derivatives of G indicated in (2.39) are easily calculated from (2.38). The results are:

$$G_\varepsilon(0, 1) = (d_{11} + \frac{1}{\alpha} d_{21})(e^{-\alpha T} - 1)$$

$$G_\lambda(0, 1) = -(e^{-\alpha T} - 1).$$

Hence, the derivative $\lambda'(0)$ is given by

(2.40) $$\lambda'(0) = d_{11} + \frac{1}{\alpha} d_{21} = \frac{1}{\alpha}\int_0^T [-g_{xx}x_1(s) - g_{xx'}x_1'(s) + f_x(s)]ds.$$

But

$$x_1(s) = C_1 + h_1(s),$$

and $x_1(s)$ has period T. Hence (2.40) simplifies to

(2.41) $$\lambda'(0) = -\frac{1}{\alpha} Tg_{xx}C_1 - \frac{1}{\alpha} g_{xx}\int_0^T h_1(s)\,ds + \frac{1}{\alpha}\int_0^T f_x(s)\,ds$$

which in turn is found to reduce to

(2.42) $$\lambda'(0) = \frac{1}{\alpha}(J_{\xi\xi}C_1 + J_{\xi\varepsilon}).$$

Since the parenthesis in (2.42) is the derivative with respect to C_1 of the left member of equation (2.21), it is not zero, since C_1 is a real simple root of (2.21).

THEOREM 2.3. *The characteristic multiplier of* (2.28) *which reduces to* +1 *when* $\varepsilon = 0$ *is greater or less than* 1 *according as*

$$\frac{\varepsilon}{\alpha}(J_{\xi\xi}C_1 + J_{\xi\varepsilon})$$

is positive or negative. Hence if α *is positive, the periodic solution is completely unstable or directly unstable according as*

$$\varepsilon(J_{\xi\xi}C_1 + J_{\xi\varepsilon})$$

is positive or negative, while if α *is negative, the periodic solution is completely stable or directly unstable according as*

$$\varepsilon(J_{\xi\xi}C_1 + J_{\xi\varepsilon})$$

is positive or negative.

Proof of Theorem 2.3. Since we have found $\lambda'(0)$ from (2.42) it follows that for small nonzero ε,

$$\lambda(\varepsilon) = 1 + \varepsilon\lambda'(0) + o(\varepsilon)$$
$$= 1 + \frac{\varepsilon}{\alpha}(J_{\xi\xi}C_1 + J_{\xi\varepsilon}) + o(\varepsilon).$$

The first assertion of the theorem follows from this. The remaining assertions follow from the fact that the second characteristic multiplier is greater or less than 1 in absolute value according as α is positive or negative. This completes the proof.

2.4. Existence and Stability in Case III. According to Section 2.1, Case III occurs if and only if $g_{x'}(0, 0) = g_x(0, 0) = 0$. The variation equation here takes the very simple form

$$y'' = 0. \tag{2.43}$$

The characteristic multipliers of (2.43) are both equal to +1, but it is not the case that all solutions of (2.43) are periodic.

We shall use the same abbreviations as were used in Section 2.3. Also we shall compute explicitly some of the partial derivatives of $F(\xi, \eta, \varepsilon)$ and $G(\xi, \eta, \varepsilon)$ at $(0, 0, 0)$.

The solutions $\phi(t)$ and $\psi(t)$ of (2.43) are

$$\phi(t) = 1, \quad \psi(t) = t.$$

The first partial derivatives of F and G at $(0, 0, 0)$ have the values

$$F_\xi = 0, \quad F_\eta = T, \quad F_\varepsilon = \int_0^T (T-s) f(s)\, ds$$

$$G_\xi = 0, \quad G_\eta = 0, \quad G_\varepsilon = \int_0^T f(s)\, ds.$$

The last two derivatives are computed as follows:

$x_\varepsilon(t)$ is that solution of $y'' = f(t)$ with $y(0) = y'(0) = 0$. Hence

$$x_\varepsilon(t) = \int_0^t (t-s) f(s)\, ds,$$

$$x_\varepsilon'(t) = \int_0^t f(s)\, ds.$$

We then find F_ε and G_ε as $x_\varepsilon(T)$ and $x_\varepsilon'(T)$ respectively.

The Jacobian of F and G with respect to ξ and η is zero, and the rank of the Jacobian matrix is 1 since $F_\eta \neq 0$. Again for $d\xi/d\varepsilon$ and $d\eta/d\varepsilon$ to exist, the matrix

$$\left\| \begin{matrix} F_\xi & F_\eta & F_\varepsilon \\ G_\xi & G_\eta & G_\varepsilon \end{matrix} \right\|$$

must have rank 1. But this requirement is equivalent to

$$\int_0^T f(s)\,ds = 0. \tag{2.44}$$

If this requirement is made, the table of first derivatives becomes

$$\begin{aligned} &F_\xi = 0 \quad F_\eta = T \quad F_\varepsilon = -\int_0^T sf(s)\,ds \\ &G_\xi = 0 \quad G_\eta = 0 \quad G_\varepsilon = 0. \end{aligned} \tag{2.45}$$

We can immediately compute $d\eta/d\varepsilon$ at $\varepsilon = 0$ from (2.45). It is given by

$$\frac{d\eta}{d\varepsilon} = -\frac{F_\varepsilon}{F_\eta} = \frac{1}{T}\int_0^T sf(s)\,ds. \tag{2.46}$$

To compute the derivative $d\xi/d\varepsilon$ at $\varepsilon = 0$, we shall need the second partial derivatives of $G(\xi, \eta, \varepsilon)$ at $(0, 0, 0)$. Since all of the first derivatives of G are zero at $(0, 0, 0)$, the second derivatives of F will not be required.

As in Section 2.3 we shall ultimately come to the function $J(\xi, \varepsilon)$ whose second derivatives at $(0, 0)$ will be used to find $d\xi/d\varepsilon$.

The second partial derivatives of G at $(0, 0, 0)$ have the following values:

$$\begin{aligned} G_{\xi\xi} &= -Tg_{xx}, \\ G_{\xi\eta} &= -\frac{T^2}{2}g_{xx} - Tg_{xx'}, \\ G_{\eta\eta} &= -\frac{T^3}{3}g_{xx} - T^2g_{xx'} - Tg_{x'x'}, \\ G_{\xi\varepsilon} &= \frac{g_{xx}}{2}\int_0^T (2Ts-s^2)f(s)\,ds + g_{xx'}\int_0^T sf(s)\,ds + \int_0^T f_x(s)\,ds, \\ G_{\eta\varepsilon} &= \frac{g_{xx}}{6}\int_0^T (3T^2s - s^3)f(s)\,ds + g_{xx'}\int_0^T Tsf(s)\,ds \\ &\quad + g_{x'x'}\int_0^T sf(s)\,ds + \int_0^T sf_x(s)\,ds + \int_0^T f_{x'}(s)\,ds, \\ G_{\varepsilon\varepsilon} &= \frac{g_{xx}}{3}\int_0^T\int_0^s (-s^3 + 3s^2u - 6Tsu)f(s)f(u)\,du\,ds \\ &\quad - g_{xx'}\left[\int_0^T sf(s)\,ds\right]^2 + 2g_{x'x'}\int_0^T\int_0^s sf(u)f(s)\,du\,ds \end{aligned} \tag{2.47}$$

$$+ 2\int_0^T\int_0^s (s-u) f_x(s) f(u)\,du\,ds + 2\int_0^T\int_0^s f_{x'}(s) f(u)\,du\,ds$$

(2.47) (continued)

$$+ 2\int_0^T f_\varepsilon(s)\,ds.$$

We give the computations for $G_{\xi\eta}$ and $G_{\xi\varepsilon}$ to illustrate typical (though simple) cases. Using the facts that $x_\xi = 1$ and $x_\eta = t$, equation (1.17) becomes

$$x''_{\xi\eta} = - g_{xx} t - g_{xx'}.$$

Integrating once, and using the fact that $x'_{\xi\eta}(0) = 0$, we find

$$x'_{\xi\eta} = - \frac{1}{2} g_{xx} t^2 - g_{xx'} t.$$

The value of $G_{\xi\eta}$ is then found from $x'_{\xi\eta}(T)$.

Using the facts that

$$x_\varepsilon(t) = \int_0^t (t-s) f(s)\,ds, \quad x'_\varepsilon(t) = \int_0^t f(s)\,ds,$$

equation (1.17d) becomes

$$x''_{\xi\varepsilon} = - g_{xx}\int_0^t (t-s) f(s)\,ds - g_{xx'}\int_0^t f(s)\,ds + f_x(t).$$

Integrating once, and using the fact that $x'_{\xi\varepsilon}(0) = 0$, we have

$$x'_{\xi\varepsilon} = - g_{xx}\int_0^t\int_0^u (u-s) f(s)\,ds\,du - g_{xx'}\int_0^t\int_0^u f(s)\,ds\,du + \int_0^t f_x(s)\,ds.$$

Interchanging the order of integration in the iterated integrals we find

$$x'_{\xi\varepsilon} = - g_{xx}\int_0^t \frac{1}{2}(t-s)^2 f(s)\,ds - g_{xx'}\int_0^t (t-s) f(s)\,ds + \int_0^t f_x(s)\,ds.$$

Setting $t = T$, and using (2.44) we obtain

$$G_{\xi\varepsilon} = \frac{g_{xx}}{2}\int_0^T (2Ts - s^2) f(s)\,ds + g_{xx'}\int_0^T s f(s)\,ds + \int_0^T f_x(s)\,ds$$

as given in (2.47).

Now define the function $H(\xi, \varepsilon)$ by the identity

$$F(\xi, H(\xi, \varepsilon), \varepsilon) \equiv 0,$$

and let

$$J(\xi, \varepsilon) = G(\xi, H(\xi, \varepsilon), \varepsilon).$$

To obtain ξ as a function of ε and to evaluate $d\xi/d\varepsilon$ at $\varepsilon = 0$, we solve the equation $J(\xi, \varepsilon) = 0$ for ξ as a function of ε, exactly as in Case II. We find that at $(0, 0)$ the first and second partial derivatives of $J(\xi, \varepsilon)$ have the values

$$J_\xi = 0, \quad J_\varepsilon = 0,$$

$$J_{\xi\xi} = -Tg_{xx},$$

$$J_{\xi\varepsilon} = \frac{g_{xx}}{2}\int_0^T (Ts-s^2) f(s)\,ds + \int_0^T f_x(s)\,ds,$$

$$\begin{aligned} J_{\varepsilon\varepsilon} = &-\frac{g_{xx}}{3}\Bigg[T\Big[\int_0^T sf(s)\,ds\Big]^2 - \frac{1}{T}\int_0^T sf(s)\,ds \int_0^T s^3 f(s)\,ds \\ &- \int_0^T\int_0^s (-s^3 + 3s^2u) f(s) f(u)\,du\,ds\Bigg] \\ &+ g_{x'x'}\Bigg[\frac{1}{T}\Big[\int_0^T sf(s)\,ds\Big]^2 + 2\int_0^T\int_0^s sf(s) f(u)\,du\,ds\Bigg] \\ &+ \frac{2}{T}\int_0^T sf(s)\,ds \int_0^T sf_x(s)\,ds + 2\int_0^T\int_0^s (s-u) f_x(s) f(u)\,du\,ds \\ &+ \frac{2}{T}\int_0^T sf(s)\,ds \int_0^T f_{x'}(s)\,ds + 2\int_0^T\int_0^s f_{x'}(s) f(u)\,du\,ds \\ &+ 2\int_0^T f_\varepsilon(s)\,ds. \end{aligned}$$

Using the values in (2.48) we have as in Case II

$$J(\xi, \varepsilon) = \frac{1}{2}J_{\xi\xi}\xi^2 + J_{\xi\varepsilon}\xi\varepsilon + \frac{1}{2}J_{\varepsilon\varepsilon}\varepsilon^2 + \text{higher order terms.}$$

If the discriminant

(2.49) $$J_{\xi\varepsilon}^2 - J_{\xi\xi}J_{\varepsilon\varepsilon}$$

is positive, the equation

(2.50) $$\frac{1}{2}J_{\xi\xi}m^2 + J_{\xi\varepsilon}m + \frac{1}{2}J_{\varepsilon\varepsilon} = 0$$

has two real simple roots if $J_{\xi\xi} \neq 0$, and one simple real root if $J_{\xi\xi} = 0$. Corresponding to each such simple real root we have a solution $\xi = \xi(\varepsilon)$ of the equation $J(\xi, \varepsilon) = 0$ with $\xi = 0$ and $d\xi/d\varepsilon = m$ at $\varepsilon = 0$.

Thus if the discriminant (2.49) is positive we are able to solve the system $F(\xi, \eta, \varepsilon) = G(\xi, \eta, \varepsilon) = 0$ for ξ and η as functions of ε near $\varepsilon = 0$, and to find the derivatives $d\xi/d\varepsilon$ and $d\eta/d\varepsilon$ at $\varepsilon = 0$. We find $d\xi/d\varepsilon$ from the roots of (2.50) and $d\eta/d\varepsilon$ from (2.46). This proves the existence of one or two periodic solutions of (1.2) in the form (1.3). The function $x_1(t)$ is completely determined from equation (2.3) and the fact that $x_1(0) = d\xi/d\varepsilon$, $x_1'(0) = d\eta/d\varepsilon$, the derivatives being evaluated at $\varepsilon = 0$.

We may also determine $x_1(t)$ from the perturbation series as in Case II. If (2.44) holds, the most general periodic solution of equation (2.3) in the present case is

(2.51) $$y = C_1 + \frac{t}{T}\int_0^T sf(s)\,ds + \int_0^t (t - s)f(s)\,ds$$

$$= C_1 + h_1(t).$$

In (2.51) C_1 is an arbitrary constant, and $h_1(t)$ is the unique periodic solution of (2.3) with $y(0) = 0$. We have thus found that

(2.52) $$x_1(t) = C_1 + h_1(t).$$

Again the derivative $x_1'(0) = h_1'(0)$ has the value given by (2.46).

The condition that the term $x_2(t)$ in the perturbation series exist is again an orthogonality condition, which furnishes, as in Case II, an equation to determine C_1. Again the equation for C_1 is equivalent to

(2.53) $$\frac{1}{2}J_{\xi\xi}C_1^2 + J_{\xi\varepsilon}C_1 + \frac{1}{2}J_{\varepsilon\varepsilon} = 0,$$

so that C_1 has the same value or values as $d\xi/d\varepsilon$, and when the discriminant is positive, C_1 will be a simple real root of (2.53).

THEOREM 2.4. *In equation* (1.2) *let* $g(x, x') \in C^2$ *and* $f(t, x, x', \varepsilon) \in C^1$. *Let equation* (1.1) *have the constant solution* $x \equiv 0$. *In addition assume that*

(2.54) $$g_{x'}(0, 0) = g_x(0, 0) = 0$$

(2.55) $$\int_0^T f(t, 0, 0, 0)\,dt = 0$$

(2.56) $$J_{\xi\varepsilon}^2 - J_{\xi\xi}J_{\varepsilon\varepsilon} > 0.$$

Then equation (1.2) *has one or two periodic solutions of the form*

(2.57) $$x = \varepsilon x_1(t) + o(\varepsilon)$$

according as $g_{xx}(0, 0) = 0$ *or* $g_{xx}(0, 0) \neq 0$. *In each case the function* $x_1(t)$ *is a uniquely determined periodic solution of*

(2.58) $$y'' = f(t, 0, 0, 0).$$

Proof of Theorem 2.4. The proof is identical with that of Theorem 2.2.

Remark: If $g_{xx}(0, 0) = 0$, and therefore in the case that $g(x, x')$ is linear, the conditions simplify. In particular condition (2.56) becomes

$$\int_0^T f_x(t, 0, 0, 0) \neq 0.$$

There is only one periodic solution in this case.

Although the existence theory was very similar in Cases II and III, the stability theory in Case III is different from that in Case II, because in Case III both characteristic multipliers of equation (2.43) are equal to +1.

By calculations similar to those given for Case II we find that

(2.59) $$\begin{aligned}\phi(T) &= 1 + \varepsilon \int_0^T (T-s)[-g_{xx}x_1(s) - g_{xx'}x_1'(s) + f_x(s)]ds + o(\varepsilon)\\ &= 1 + \varepsilon d_{11} + o(\varepsilon),\end{aligned}$$

(2.60) $$\begin{aligned}\phi'(T) &= \varepsilon \int_0^T [-g_{xx}x_1(s) - g_{xx'}x_1'(s) + f_x(s)]ds + o(\varepsilon)\\ &= \varepsilon d_{21} + o(\varepsilon)\end{aligned}$$

(2.61) $$\psi(T) = T + \varepsilon d_{12} + o(\varepsilon)$$

(2.62) $$\begin{aligned}\psi'(T) &= 1 + \varepsilon \int_0^T \{[-g_{xx}x_1(s) - g_{xx'}x_1'(s) + f_x(s)]s\\ &\qquad + [-g_{xx'}x_1(s) - g_{x'x'}x_1'(s) + f_{x'}(s)]\}ds + o(\varepsilon)\\ &= 1 + \varepsilon d_{22} + o(\varepsilon).\end{aligned}$$

The value of d_{12} is not important for the theory. The characteristic multipliers are the roots of

$$(2.63) \qquad G(\varepsilon,\lambda) = \det \begin{Vmatrix} 1+\varepsilon d_{11} + o(\varepsilon) - \lambda & T + \varepsilon d_{12} + o(\varepsilon) \\ \varepsilon d_{21} + o(\varepsilon) & 1 + \varepsilon d_{22} + o(\varepsilon) - \lambda \end{Vmatrix}$$

$$= \lambda^2 - \lambda(2 + \varepsilon(d_{11} + d_{22}) + o(\varepsilon))$$

$$+ (1 + \varepsilon(d_{11} + d_{22} - Td_{21}) + o(\varepsilon)) = 0.$$

The roots of (2.63) are

$$(2.64) \qquad \lambda = 1 + \frac{1}{2}\varepsilon(d_{11} + d_{22}) \pm \sqrt{\varepsilon T d_{21}} + o(\varepsilon).$$

Examination of (2.64) shows that the two characteristic multipliers are real and separated by 1 for small ε if $\varepsilon T d_{21}$ is positive. When $\varepsilon T d_{21}$ is negative, the roots are complex for small ε. They are less than 1 in absolute value if $d_{11} + d_{22} - Td_{21}$ is negative, and greater than 1 in absolute value if $d_{11} + d_{22} - Td_{21}$ is positive.

From (2.60),

$$d_{21} = \int_0^T [-g_{xx} x_1(s) - g_{xx'} x_1'(s) + f_x(s)]\,ds.$$

But using the fact that $x_1(s) = C_1 + h_1(s)$, this is found to reduce to

$$d_{21} = J_{\xi\xi} C_1 + J_{\xi\varepsilon}.$$

Consequently, d_{21} is never zero, since it is the derivative with respect to C_1 of the left member of equation (2.53) of which C_1 is a simple real root.

From (2.59), (2.60), and (2.62)

$$d_{11} + d_{22} - Td_{21} = \int_0^T [-g_{xx'} x_1(s) - g_{x'x'} x_1'(s) + f_{x'}(s)]\,ds$$

$$= -Tg_{xx'} C_1 - \frac{g_{xx'}}{2} \int_0^T (Ts - s^2) f(s)\,ds + \int_0^T f_{x'}(s)\,ds.$$

THEOREM 2.5. *The stability of a periodic solution as found in Theorem* 2.4 *is determined as follows:*

(a) *The solution is directly unstable if*

$$\varepsilon T\left(-Tg_{xx}C_1 - \frac{g_{xx}}{2}\int_0^T (Ts - s^2) f(s)\,ds + \int_0^T f_x(s)\,ds\right) \tag{2.65}$$

is positive.

(b) *If* (2.65) *is negative, the solution is completely stable or completely unstable according as*

$$\varepsilon\left(-Tg_{xx'}C_1 - \frac{g_{xx'}}{2}\int_0^T (Ts - s^2) f(s)\,ds + \int_0^T f_{x'}(s)\,ds\right) \tag{2.66}$$

is negative or positive.

Proof of Theorem 2.5. The proof is contained in the discussion preceding the statement of the theorem. The only remark to make is that (2.65) is obtained by writing explicitly the expression for $J_{\xi\xi}C_1 + J_{\xi\varepsilon}$.

Remarks:

1. If there are two periodic solutions, (2.65) always has opposite signs for the two solutions, so that one is always directly unstable.

2. In the case that g is linear, the stability depends only on the signs of $\int_0^T f_x(s)\,ds$ and $\int_0^T f_{x'}(s)\,ds$.

2.5. Existence and Stability in Case IV. We know from Section 2.1 that Case IV occurs if and only if $g_{x'}(0, 0) = 0$ and $g_x(0, 0) = 4n^2\pi^2T^{-2}$ for some positive integer n. The variation equation is then

$$y'' + \frac{4n^2\pi^2}{T^2}\, y = 0. \tag{2.67}$$

All solutions of (2.67) have period T, so that both characteristic multipliers are equal to +1.

We continue to use the abbreviations introduced in Section 2.3. It turns out that the partial derivatives of F and G at $(0, 0, 0)$ can be expressed in terms of the Fourier coefficients of $f(t, 0, 0, 0)$, $f_x(t, 0, 0, 0)$, and $f_{x'}(t, 0, 0, 0)$. Let the Fourier series for these three functions be

$$f(t, 0, 0, 0) \sim \frac{1}{2} a_o + \sum_{n=1}^{\infty} a_n \cos \frac{2n\pi t}{T} + b_n \sin \frac{2n\pi t}{T} ; \tag{2.68}$$

$$f_x(t, 0, 0, 0) \sim \frac{1}{2} a_{ox} + \sum_{n=1}^{\infty} a_{nx} \cos \frac{2n\pi t}{T} + b_{nx} \sin \frac{2n\pi t}{T} ; \tag{2.69}$$

$$f_{x'}(t, 0, 0, 0) \sim \frac{1}{2} a_{ox} + \sum_{n=1}^{\infty} a_{nx'} \cos \frac{2n\pi t}{T} + b_{nx'} \sin \frac{2n\pi t}{T} . \tag{2.70}$$

The solutions $\phi(t)$ and $\psi(t)$ of (2.67) are

$$\phi(t) = \cos \frac{2n\pi t}{T}, \quad \psi(t) = \frac{T}{2n\pi} \sin \frac{2n\pi t}{T} .$$

The first partial derivatives of F and G at $(0, 0, 0)$ have the values

$$F_\xi = 0 \qquad F_\eta = 0 \qquad F_\varepsilon = -\frac{T^2}{4n\pi} b_n$$

$$G_\xi = 0 \qquad G_\eta = 0 \qquad G_\varepsilon = \frac{T}{2} a_n$$

where a_n and b_n are coefficients from (2.68).

The Jacobian of F and G with respect to ξ and η is zero, and the rank of the matrix is zero. Therefore for $d\xi/d\varepsilon$ and $d\eta/d\varepsilon$ to exist, the rank of

$$\left\| \begin{matrix} F_\xi & F_\eta & F_\varepsilon \\ G_\xi & G_\eta & G_\varepsilon \end{matrix} \right\|$$

must be zero. But this is equivalent to

$$a_n = b_n = 0. \tag{2.71}$$

With this requirement satisfied, all six first partial derivatives of F and G are zero. Hence to find $d\xi/d\varepsilon$ and $d\eta/d\varepsilon$, we shall need the second partial derivatives of F and G at $(0, 0, 0)$.

The second partial derivatives of F and G at $(0, 0, 0)$ have the values:

$$F_{\xi\xi} = G_{\xi\xi} = F_{\xi\eta} = G_{\xi\eta} = F_{\eta\eta} = G_{\eta\eta} = 0,$$

$$F_{\xi\varepsilon} = -\frac{T^4 g_{xx}}{96n^3\pi^3} b_{2n} + \frac{T^3 g_{xx'}}{48n^2\pi^2}(a_{2n} - 3a_o) - \frac{T^2 g_{x'x'}}{12n\pi} b_{2n}$$

$$- \frac{T^2}{8n\pi} b_{2nx} - \frac{T}{4}(a_{2nx'} - a_{ox'}),$$

$$G_{\xi\varepsilon} = \frac{T^3 g_{xx}}{4\varepsilon n^2\pi^2}(a_{2n} - 3a_o) + \frac{T^2 g_{xx'}}{24n\pi} b_{2n} + \frac{T g_{x'x'}}{6} a_{2n}$$

(2.72)
(continued)

$$+ \frac{T}{4}(a_{2nx} + a_{ox}) - \frac{n\pi}{2} b_{2nx'},$$

$$F_{\eta\varepsilon} = \frac{T^5 g_{xx}}{192n^4\pi^4}(a_{2n} + 3a_o) + \frac{T^4 g_{xx'}}{96n^3\pi^3} b_{2n} + \frac{T^3 g_{x'x'}}{24n^2\pi^2} a_{2n}$$

$$+ \frac{T^3}{16n^2\pi^2}(a_{2nx} - a_{ox}) - \frac{T^2}{8n\pi} b_{2nx'},$$

$$G_{\eta\varepsilon} = \frac{T^4 g_{xx}}{96n^3\pi^3} b_{2n} - \frac{T^3 g_{xx'}}{48n^2\pi^2}(a_{2n} + 3a_o) + \frac{T^2 g_{x'x'}}{12n\pi} b_{2n}$$

$$+ \frac{T^2}{8n\pi} b_{2nx} + \frac{T}{4}(a_{2nx'} + a_{ox'})$$

The expressions for $F_{\varepsilon\varepsilon}$ and $G_{\varepsilon\varepsilon}$ are extremely long and complicated, and will not be given.

We can now write expressions for $F(\xi, \eta, \varepsilon)$ and $G(\xi, \eta, \varepsilon)$ valid near $(0, 0, 0)$:

$$F(\xi, \eta, \varepsilon) = F_{\xi\varepsilon}\xi\varepsilon + F_{\eta\varepsilon}\eta\varepsilon + \frac{1}{2}F_{\varepsilon\varepsilon}\varepsilon^2 + \text{higher order terms}$$

$$G(\xi, \eta, \varepsilon) = G_{\xi\varepsilon}\xi\varepsilon + G_{\eta\varepsilon}\eta\varepsilon + \frac{1}{2}G_{\varepsilon\varepsilon}\varepsilon^2 + \text{higher order terms.}$$

Now by a modification of the implicit function theorem as given in [5], Chapter IV, we have that if the system

(2.73)

$$F_{\xi\varepsilon}\alpha + F_{\eta\varepsilon}\beta + \frac{1}{2}F_{\varepsilon\varepsilon} = 0$$

$$G_{\xi\varepsilon}\alpha + G_{\eta\varepsilon}\beta + \frac{1}{2}G_{\varepsilon\varepsilon} = 0$$

has a unique solution (i. e. if $F_{\xi\varepsilon}G_{\eta\varepsilon} - F_{\eta\varepsilon}G_{\xi\varepsilon} \neq 0$), then we can find ξ and η as functions of ε for small ε with $\xi = \eta = 0$ at $\varepsilon = 0$ and $d\xi/d\varepsilon = \alpha$, $d\eta/d\varepsilon = \beta$ at $\varepsilon = 0$. Thus assuming the determinant is not zero, we have the existence proof for the periodic solution.

We can also determine $d\xi/d\varepsilon$ and $d\eta/d\varepsilon$ formally from the perturbation series. Since $x_1(t)$ is a solution of

$$y'' + \frac{4n^2\pi^2}{T^2} y = f(t, 0, 0, 0), \tag{2.74}$$

and since every solution of (2.74) has period T in view of (2.71),

$$\begin{aligned} x_1(t) &= C_1 \cos\frac{2n\pi t}{T} + C_2 \frac{T}{2n\pi}\sin\frac{2n\pi t}{T} \\ &\quad + \int_0^t \frac{T}{2n\pi}\sin\frac{2n\pi}{T}(t-s) f(s)\,ds. \end{aligned} \tag{2.75}$$

Now when (2.75) is used for x_1 in the orthogonality conditions for the existence of a periodic $x_2(t)$ in the perturbation series, the conditions furnish as equations for C_1 and C_2

$$\begin{aligned} F_{\xi\varepsilon}C_1 + F_{\eta\varepsilon}C_2 + \tfrac{1}{2}F_{\varepsilon\varepsilon} &= 0 \\ G_{\xi\varepsilon}C_1 + G_{\eta\varepsilon}C_2 + \tfrac{1}{2}G_{\varepsilon\varepsilon} &= 0 \end{aligned} \tag{2.76}$$

which is the same as (2.73), so that $C_1 = x_1(0)$ and $C_2 = x_1'(0)$ have the same values as found for $d\xi/d\varepsilon$ and $d\eta/d\varepsilon$.

THEOREM 2.6. *In equation* (1.2) *let* $g(x, x') \in C^2$ *and* $f(t, x, x', \varepsilon) \in C^1$. *Let equation* (1.1) *have the constant solution* $x \equiv 0$. *In addition assume that*

$$g_{x'}(0, 0) = 0 \tag{2.77}$$

$$g_x(0, 0) = \frac{4n^2\pi^2}{T^2} \tag{2.78}$$

for some positive integer n.

(2.79) *The Fourier coefficients* a_n *and* b_n *in* (2.68) *are both zero.*

$$F_{\xi\varepsilon}G_{\eta\varepsilon} - F_{\eta\varepsilon}G_{\xi\varepsilon} \neq 0. \tag{2.80}$$

Then there exists a unique periodic solution of (1.2) *of the form*

(2.81) $$x = \varepsilon x_1(t) + o(\varepsilon)$$

where $x_1(t)$ *is given by* (2.75) *with* C_1 *and* C_2 *determined by* (2.76).

Proof of Theorem 2.6. The proof of the theorem is contained in the discussion which precedes it.

The stability investigation in Case IV is somewhat simpler than in Cases II and III. Calculations similar to those made in earlier cases show that

(2.82) $$\phi(T) = 1 + \varepsilon F_{\xi\varepsilon} + o(\varepsilon)$$

(2.83) $$\phi'(T) = \varepsilon G_{\xi\varepsilon} + o(\varepsilon)$$

(2.84) $$\psi(T) = \varepsilon F_{\eta\varepsilon} + o(\varepsilon)$$

(2.85) $$\psi'(T) = 1 + \varepsilon G_{\eta\varepsilon} + o(\varepsilon)$$

Thus the characteristic multipliers are the eigenvalues of

$$I + \varepsilon \left\| \begin{matrix} F_{\xi\varepsilon} & F_{\eta\varepsilon} \\ G_{\xi\varepsilon} & G_{\eta\varepsilon} \end{matrix} \right\| + o(\varepsilon)$$

and so are given by

$$\lambda_1 = 1 + \varepsilon\mu_1 + o(\varepsilon), \quad \lambda_2 = 1 + \varepsilon\mu_2 + o(\varepsilon)$$

where μ_1 and μ_2 are the eigenvalues of

$$\left\| \begin{matrix} F_{\xi\varepsilon} & F_{\eta\varepsilon} \\ G_{\xi\varepsilon} & G_{\eta\varepsilon} \end{matrix} \right\|$$

The characteristic multipliers are real and separated by +1 when μ_1 and μ_2 have opposite sign, i.e. when $F_{\xi\varepsilon}G_{\eta\varepsilon} - F_{\eta\varepsilon}G_{\xi\varepsilon} < 0$. When $F_{\xi\varepsilon}G_{\eta\varepsilon} - F_{\eta\varepsilon}G_{\xi\varepsilon} > 0$, the characteristic multipliers are less than 1 in absolute value if the μ's have negative real part, and are greater than 1 in absolute value if the μ's have positive real part. The determining quantity in this case is $F_{\xi\varepsilon} + G_{\eta\varepsilon}$, which is equal, by (2.72) to

$$-\frac{T^3 g_{xx'}}{8n^2\pi^2} a_o + \frac{T}{2} a_{ox'} = -\frac{T^2 g_{xx'}}{4n^2\pi^2} \int_0^T f(s)\,ds + \int_0^T f_{x'}(s)\,ds.$$

THEOREM 2.7. *The stability of the periodic solution found in Theorem 2.6 is as follows:*

(a) *The solution is directly unstable if*

(2.86) $$F_{\xi\varepsilon} G_{\eta\varepsilon} - F_{\eta\varepsilon} G_{\xi\varepsilon}$$

is negative.

(b) *If* (2.86) *is positive, the solution is completely stable or completely unstable according as*

(2.87) $$\varepsilon \left[-\frac{T^2 g_{xx'}}{4n^2\pi^2} \int_0^T f(s)\,ds + \int_0^T f_{x'}(s)\,ds \right]$$

is negative or positive.

Proof: The theorem is proved by the discussion which precedes it.

2.6. The Case that $f(t, x, x', \varepsilon)$ is independent of t. As a rule in the study of perturbations of periodic solutions, when the forcing term $f(t, x, x', \varepsilon)$ is independent of t, the situation is more complicated, because the period of the new periodic solution may well be different from the period of the unperturbed solution. However, when the solution that is perturbed is a constant, these difficulties are not present. The theory of Part II of this paper applies, and in the present section we give the results.

The results are extremely simple. In fact, every periodic solution shown to exist by our theorems is identically constant, with the stability as given by our results.

THEOREM 2.8. *In the equation*

(2.88) $$x'' + g(x, x') = \varepsilon f(x, x', \varepsilon)$$

let $g(x, x') \in C^2$ *and* $f(x, x', \varepsilon) \in C^1$. *Let* $x \equiv 0$ *be a solution when* $\varepsilon = 0$.

Then if $g_x(0, 0) \neq 0$, (2.88) *has a unique constant solution of the form*

(2.89) $$x = \varepsilon x_1 + o(\varepsilon)$$

where x_1 *is the constant* $f(0,0,0)/g_x(0,0)$.

If $g_x(0,0) = 0$, *there is no solution of the form* (2.89) *unless* $f(0,0,0) = 0$ *also. In this case, there are one or two solutions of the form* (2.89) *where* x_1 *is a root of the equation*

$$-\frac{1}{2}g_{xx}m^2 + f_x m + f_\varepsilon = 0 \tag{2.90}$$

provided that the discriminant of (2.90) *is positive.*

Proof of Theorem 2.8. If x is a constant solution of (2.88) it satisfies the equation

$$H(x,\varepsilon) \equiv \varepsilon f(x,0,\varepsilon) - g(x,0) = 0. \tag{2.91}$$

Clearly $H(0,0) = 0$, $H_x(0,0) = -g_x(0,0)$, and $H_\varepsilon(0,0) = f(0,0,0)$. Hence by the implicit function theorem we can solve for x as a function of ε if $g_x(0,0) \neq 0$, and $dx/d\varepsilon = f(0,0,0)/g_x(0,0)$ at $\varepsilon = 0$.

If $g_x(0,0) = 0$ and $f(0,0,0) \neq 0$ no solution of $H = 0$ of the form (2.89) exists. If $f(0,0,0) = 0$ also, we compute the second partial derivatives of H and find

$$H_{xx} = -g_{xx}(0,0), \quad H_{x\varepsilon} = f_x(0,0,0), \quad H_{\varepsilon\varepsilon} = 2f_\varepsilon(0,0,0).$$

Hence near $(0,0)$, we have

$$H(x,\varepsilon) = -\frac{1}{2}g_{xx}x^2 + f_x x\varepsilon + f_\varepsilon \varepsilon^2 + \text{higher order terms},$$

so that by the modified implicit function theorem [9], Lemma 8, we may solve for x as a function of ε provided that the discriminant $f_x^2 + 2f_\varepsilon g_{xx}$ is positive. This completes the proof of Theorem 2.8.

We can now apply Theorem 2.8 and discuss the various cases studied earlier. In Cases I and IV, where $g_x(0,0) \neq 0$, the theory gave a unique periodic solution which can only be the constant solution found in Theorem 2.8. Indeed it can be verified that in Cases I and IV the formulae for $d\xi/d\varepsilon$ and $d\eta/d\varepsilon$ give $f(0,0,0)/g_x(0,0)$ and 0 respectively.

In case IV, which occurs only when $g_x(0,0)$ is positive, $g_{x'}(0,0) = 0$, and the period T is related to $g_x(0,0)$ by

$$T^2 g_x(0,0) = 4n^2\pi^2$$

for some positive integer n, the determinant (2.80) has the value

$$\frac{T^2}{4}\left(f_{x'}(0,0,0) - \frac{g_{xx'}(0,0)}{g_x(0,0)} f(0,0,0)\right)^2 + \frac{T^4}{16n^2\pi^2}\left(f_x(0,0,0) - \frac{g_{xx}(0,0)}{g_x(0,0)} f(0,0,0)\right)^2$$

Thus it is almost always positive. Hence the constant solution is never directly unstable in Case IV. It turns out to be completely stable or completely unstable according as

$$\varepsilon\left(f_{x'}(0,0,0) - \frac{g_{xx'}(0,0)}{g_x(0,0)} f(0,0,0)\right) \tag{2.92}$$

is negative or positive, by (2.87).

In Cases II and III where $g_x(0,0) = 0$, the condition (2.11) or (2.44) gives $f(0,0,0) = 0$. Then the equation for C_1 (2.21) or (2.53) turns out to be equivalent to (2.90), so that the periodic solutions found in Cases II and III are again the constant solutions found in Theorem 2.8.

2.7. Concluding Remarks. In conclusion we point out the cases that all under Part II of this paper which are not covered.

In Case II we have not considered the cases in which the discriminant (2.25) is nonpositive. This would involve nonreal solutions, or would require the study of higher derivatives of the functions F and G.

In Case III we have not considered the cases in which the discriminant (2.56) is nonpositive. This would involve nonreal solutions, or would require the study of higher derivatives of the functions F and G. Also in the stability theorem we have not considered the case in which (2.65) is negative and (2.66) is zero. It is not possible to distinguish among complete stability, quasi-stability, and complete instability without going to higher derivatives of F and G.

In Case IV we have not considered the cases in which the determinant (2.80) is zero. This would require a study of higher order derivatives of F and G. In the stability theorem we have not considered the case in which (2.86) is positive and (2.87) is zero. Again it requires higher derivatives of F and G to distinguish among the various possibilities for stability.

It remains to determine the stability of the periodic solution found in Theorem 2.1 in the case mentioned in the first remark following the theorem.

THEOREM 2.9. *In equation* (1.2) *let* $g(x, x') \in C^2$ *and* $f(t, x, x', \varepsilon) \in C^1$. *Let* $x \equiv 0$ *be a solution when* $\varepsilon = 0$. *Let* $g_{x'}(0, 0) = 0$; $g_x(0, 0) = k^2 > 0$, *where* $\cos kT \neq 1$. *Then:*

(a) *If* $\cos kT \neq -1$ *as well, the periodic solution found in Theorem* 2.1 *is completely stable or completely unstable according as*

$$\varepsilon \left[-\frac{g_{xx'}}{k^2} \int_0^T f(s)\,ds + \int_0^T f_{x'}(s)\,ds \right] \tag{2.93}$$

is negative or positive.

(b) *If* $\cos kT = -1$, *so that* $k^2 = (2n-1)^2 \pi^2 T^{-2}$ *for some positive integer* n, *the solution is inversely unstable if*

$$d_{11} d_{22} - d_{21} d_{12} \tag{2.94}$$

is negative. If (2.94) *is positive, the solution is completely stable or completely unstable according as* (2.93) *is negative or positive.*

Note. The symbols in (2.94) are defined in the course of the proof. We have used the abbreviations introduced in Case II.

Remark. We do not cover the cases where (2.94) is zero, nor when (2.93) is zero.

The similarity of (2.93) to (2.87) and (2.92) should be noted.

Proof of Theorem 2.9. We begin by proving assertion (a). The variation equation corresponding to the periodic solution can be written

$$\begin{aligned} y'' + k^2 y = \varepsilon \{ & [-g_{xx} x_1(t) - g_{xx'} x_1'(t) + f_x(t)] y \\ & + [-g_{xx'} x_1(t) - g_{x'x'} x_1'(t) + f_{x'}(t)] y' \} + o(\varepsilon). \end{aligned} \tag{2.94}$$

Solving by successive approximations we find as earlier

$$\phi(T) = \cos kT + \varepsilon d_{11} + o(\varepsilon), \quad \psi(T) = \frac{\sin kT}{k} + \varepsilon d_{12} + o(\varepsilon),$$

(2.95)

$$\phi'(T) = -k \sin kT + \varepsilon d_{21} + o(\varepsilon), \quad \psi'(T) = \cos kT + \varepsilon d_{22} + o(\varepsilon).$$

The explicit expressions for d_{11}, d_{12}, d_{21}, and d_{22} are not needed for assertion (a), and so we do not write them out. The one fact that will be needed is that

$$(d_{11} + d_{22}) \cos kT + (kd_{12} - \frac{1}{k} d_{21}) \sin kT$$

(2.96)

$$= \int_0^T [-g_{xx'} x_1(s) - g_{x'x'} x_1'(s) + f_{x'}(s)]\, ds.$$

The characteristic multipliers of (2.94) are the eigenvalues of the matrix

(2.97) $$\left\| \begin{matrix} \phi(T) & \psi(T) \\ \phi'(T) & \psi'(T) \end{matrix} \right\|$$

When $\varepsilon = 0$, they are $e^{\pm ikT}$, and in case (a) are not real. Hence for small nonzero ε they remain nonreal, and conjugate complex. Hence to determine their absolute values we need only study their product. But the product is just the determinant of the matrix (2.97), and using (2.95) this is

$$1 + \varepsilon(d_{11} \cos kT + kd_{12} \sin kT - \frac{1}{k} d_{21} \sin kT + d_{22} \cos kT) + o(\varepsilon)$$

$$= 1 + \varepsilon \{\int_0^T [-g_{xx'} x_1(s) - g_{x'x'} x_1'(s) + f_{x'}(s)]\, ds\} + o(\varepsilon),$$

by (2.96). Now $x_1(t)$ has period T. Hence $\int_0^T x_1'(s)\, ds = 0$. Also, $x_1(t)$ satisfies

$$x_1'' + k^2 x_1 = f(t).$$

Integrating this from 0 to T, we obtain

$$k^2 \int_0^T x_1(s)\, ds = \int_0^T f(s)\, ds.$$

Therefore the product of the characteristic multipliers is

$$1 + \varepsilon \left[- \frac{g_{xx'}}{k^2} \int_0^T f(s)\, ds + \int_0^T f_{x'}(s)\, ds \right] + o(\varepsilon),$$

and this implies assertion (a) of the theorem.

We now prove assertion (b). Here we assume $k = (2n-1)\pi/T$ for some positive integer n. We shall have occasion to use the Fourier coefficients from (2.68), (2.69) and (2.70). Solving (2.94) by successive approximations we find

$$\text{(2.98)}\qquad \begin{aligned} \phi(T) &= -1 + \varepsilon d_{11} + o(\varepsilon), & \psi(T) &= \varepsilon d_{12} + o(\varepsilon) \\ \phi'(T) &= \varepsilon d_{21} + o(\varepsilon), & \psi'(T) &= -1 + \varepsilon d_{22} + o(\varepsilon) \end{aligned}$$

where

$$d_{11} = \frac{Tg_{xx}}{12k^3} b_{2n-1} - \frac{Tg_{xx'}}{12k^2}(a_{2n-1} - 3a_o) + \frac{Tg_{x'x'}}{6k} b_{2n-1}$$
$$+ \frac{T}{4k} b_{2n-1,x} + \frac{T}{4}(a_{2n-1,x'} - a_{ox'})$$

$$d_{12} = -\frac{Tg_{xx}}{12k^4}(a_{2n-1} + 3a_o) - \frac{Tg_{xx'}}{12k^3} b_{2n-1} - \frac{Tg_{x'x'}}{6k^2} a_{2n-1}$$
$$- \frac{T}{4k^2}(a_{2n-1,x} - a_{ox}) + \frac{T}{4k} b_{2n-1,x'}$$

$$d_{21} = -\frac{Tg_{xx}}{12k^2}(a_{2n-1} - 3a_o) - \frac{Tg_{xx'}}{12k} b_{2n-1} - \frac{Tg_{x'x'}}{6} a_{2n-1}$$
$$- \frac{T}{4}(a_{2n-1,x} + a_{ox}) + \frac{kT}{4} b_{2n-1,x'}$$

$$d_{22} = -\frac{Tg_{xx}}{12k^3} b_{2n-1} + \frac{Tg_{xx'}}{12k^2}(a_{2n-1} + 3a_o) - \frac{Tg_{x'x'}}{6k} b_{2n-1}$$
$$- \frac{T}{4k} b_{2n-1,x} - \frac{T}{4}(a_{2n-1,x'} + a_{ox'}).$$

Now in case (b), the matrix (2.97) is

$$-I + \varepsilon \left\| \begin{matrix} d_{11} & d_{12} \\ d_{21} & d_{22} \end{matrix} \right\| + o(\varepsilon)$$

so that the characteristic multipliers can be written

$$\lambda_1 = -1 + \varepsilon\mu_1 + o(\varepsilon) \quad \text{and} \quad \lambda_2 = -1 + \varepsilon\mu_2 + o(\varepsilon)$$

where μ_1 and μ_2 are the eigenvalues of the matrix

$$\text{(2.99)}\qquad \left\| \begin{matrix} d_{11} & d_{12} \\ d_{21} & d_{22} \end{matrix} \right\|$$

Now if the determinant of (2.99) is negative, the μ's are real with opposite signs, so that the λ's are real and separated by -1. This produces inverse instability. In case the determinant of (2.99) is positive, the real parts of the μ's have the same sign. This sign is the sign of the trace $(d_{11} + d_{22})$. Hence if $\varepsilon(d_{11} + d_{22})$ is positive, the λ's have real part greater than -1 and consequently have absolute value less than 1. If $\varepsilon(d_{11} + d_{22})$ is negative, the λ's have real part less than -1 and consequently have absolute value greater than 1. Thus when (2.94) is positive, we have complete stability or complete instability according as $\varepsilon(d_{11} + d_{22})$ is positive or negative. But

$$\begin{aligned} \varepsilon(d_{11} + d_{22}) &= \frac{\varepsilon T}{2} \frac{g_{xx'}}{k^2} a_o - a_{ox'} \\ &= \left[\frac{-g_{xx'}}{k^2} \int_0^T f(s)\,ds - \int_0^T f_{x'}(s)\,ds \right] \end{aligned}$$

and since this is the negative of (2.93), assertion (b) is proved. This completes the proof of Theorem 2.9.

As an example of inverse instability, the periodic solution of

$$x'' + (1 + \varepsilon \cos 2t)x = \varepsilon f(t)$$

where $f(t)$ is any function of period π, is inversely unstable for small nonzero ε.

PART III: PERTURBATION OF A NONCONSTANT SOLUTION

3.1. Introduction. In this part of the paper we assume that the periodic solution, $x_0(t)$, of (1.1) is not identically constant. Let its least period be L_0. We choose the time origin at a point where $x_0'(t) = 0$. Let the corresponding value of x_0 be A, so that $x_0(0) = A$, and $x_0'(0) = 0$. This will imply that $g(A, 0) \neq 0$.

It will be necessary to assume that the least period in t of $f(t, x, x', \varepsilon)$, which we denote by T, is a rational multiple of L_0. We denote the least common multiple of T and L_0 by L.

We shall study periodic solutions of (1.2) having the form

$$x = x_0(t) + \varepsilon x_1(t) + o(\varepsilon). \tag{3.1}$$

When we have occasion to study periodic solutions of (1.2) of this type which reduce for $\varepsilon = 0$ to a translation of $x_0(t)$, $x_0(t+\tau)$, we do it by studying the equivalent problem obtained by replacing $f(t, x, x', \varepsilon)$ in (1.2) by $f(t-\tau, x, x', \varepsilon)$.

The associated variational equation is

$$L(y) \equiv y'' + g_{x'}(x_0(t), x_0'(t))y' + g_x(x_0(t), x_0'(t))y = 0. \tag{3.2}$$

Equation (3.2) has the nontrivial periodic solution $y = x_0'(t)$, which has period L_0 and hence also has period L. The characteristic multipliers of (3.2) will be defined considering it as an equation whose coefficients have period L. Namely, the characteristic multipliers will be the eigenvalues of the matrix

$$\begin{Vmatrix} \phi(L) & \psi(L) \\ \phi'(L) & \psi'(L) \end{Vmatrix}$$

Because of the periodic solution $y = x_0(t)$, one characteristic multiplier is +1. As we saw in Part I, the second characteristic multiplier has the value

$$e^{-\int_0^L g_{x'}(x_0(t), x_0'(t))\,dt}$$

Of the four cases mentioned in Part I, only three can occur when $x_0(t)$ is nonconstant. Case II occurs when only one of the characteristic multipliers is equal to +1, hence Case II occurs if and only if

$$\int_0^L g_{x'}(x_0(t), x_0'(t))\,dt \neq 0.$$

Cases III and IV occur when the second characteristic multiplier is equal to +1, namely when

$$\int_0^L g_{x'}(x_0(t), x_0'(t))\,dt = 0.$$

In Case III the only periodic solutions of (3.2) are $x_0'(t)$ and its constant multiples. In Case IV, every solution of (3.2) is periodic with period L (and least period L_0).

As in Part II we shall omit the computations of the partial derivatives of F and G at $(0, 0, 0)$ in general. Some sample computations will be given in Case III.

The results of this part overlap somewhat with those of Malkin [10]. His results in general are the same as those we obtain which use only the second partial derivatives of F and G. We go beyond this point and obtain the results which require the third partial derivatives of F and G. Our classification of cases is also different from Malkin's. His fundamental classification is according to the number of parameters on which the periodic solution of the autonomous equation can depend. When there is an isolated periodic solution of the autonomous equation, it and its translations form a one-parameter family of periodic solutions. If, as in the linear case, all solutions of the autonomous equation have the same period, (or all in some neighborhood of a particular solution have the same period), there is a two-parameter family of periodic solutions of the autonomous equation, the variation equation does indeed have all its solutions periodic, and our Case IV occurs. However, it is possible for Case IV to occur even if there is an isolated periodic solution of the autonomous equation, as we shall point out. This fact is apparently overlooked in Malkin's book as also in the paper of Kac [6].

3.2. <u>Existence and Stability in Case II.</u> As we saw in Section 3.1, Case II occurs if and only if

$$\frac{1}{L}\int_0^L g_{x'}(x_0(t), x_0'(t))\,dt \neq 0. \tag{3.3}$$

We denote by α the value of the left member of (3.3). The characteristic multipliers of the variational equation (3.2) are +1 and $e^{-\alpha L}$.

The results of this section can all be obtained by specializing the results of [9] to the case $n = 2$ and making appropriate changes of notation. In the present treatment we obtain them directly using the techniques introduced here.

For the purposes of this section we shall need a particular fundamental set of solutions of (3.2). The first of these is $x_0'(t)$, which we shall denote by $p(t)$. The function $p(t)$ has period L, and satisfies

$$(3.4)\qquad \begin{aligned} p(0) &= p(L) = x_0'(0) = 0, \\ p'(0) &= p'(L) = x_0''(0) = -g(A, 0). \end{aligned}$$

By Floquet theory (cf. [3], Chap. 3), (3.2) has a solution of the form $r(t)e^{-\alpha t}$, where $r(t)$ has period L. As our second fundamental solution we choose an appropriate constant multiple of this solution, which we call $q(t)$. We require that $q(0) = 1/g(A, 0)$. The form and initial value of $q(t)$ together uniquely determine $q'(0)$, which cannot be expressed in terms of quantities already introduced. The function $q(t)$ satisfies

$$(3.5)\qquad \begin{aligned} q(0) &= \frac{1}{g(A, 0)}, \quad q(L) = \frac{e^{-\alpha L}}{g(A,0)} \\ q'(L) &= e^{-\alpha L} q'(0). \end{aligned}$$

Because of the selection of initial values, the Wronskian

$$W(t) = p(t)q'(t) - p'(t)q(t)$$

satisfies $W(0) = 1$. Since $W(t)$ is known to satisfy

$$(3.6)\qquad W'(t) + g_{x'}(x_0(t), x_0'(t))q(t) = 0$$

it follows that

$$W(t) = e^{-\int_0^t g_{x'}(x_0(s), x_0'(s))\,ds}$$

Hence $W(L) = e^{-\alpha L}$, and $W(t)$ is a periodic function multiplied by $e^{-\alpha t}$. As a result, the quotient $q(t)/W(t)$ is a periodic function.

The solutions $\phi(t)$ and $\psi(t)$ of (3.2) (which satisfy $\phi(0) = \psi'(0) = 1$,

$\phi'(0) = \psi(0) = 0)$ are given by

(3.7)
$$\phi(t) = g(A, 0)\,q(t) + q'(0)\,p(t),$$
$$\psi(t) = -\frac{p(t)}{g(A, 0)}.$$

The first partial derivatives of F and G at $(0, 0, 0)$ have the values

$$F_\xi = e^{-\alpha L} - 1$$

$$G_\xi = g(A, 0)\,q'(0)\,(e^{-\alpha L} - 1)$$

$$F_\eta = 0$$

$$G_\eta = 0$$

$$F_\varepsilon = \frac{e^{-\alpha L}}{g(A, 0)} \int_0^L \frac{p(s)}{W(s)} f(s, x_0(s), x_0'(s), 0)\,ds$$

$$G_\varepsilon = e^{-\alpha L} q'(0) \int_0^L \frac{p(s)}{W(s)} f(s, x_0(s), x_0'(s), 0)\,ds$$
$$+ g(A, 0) \int_0^L \frac{q(s)}{W(s)} f(s, x_0(s), x_0'(s), 0)\,ds.$$

The Jacobian of F and G with respect to ξ and η is zero, and the rank of the Jacobian matrix is 1, since $F_\xi \neq 0$. As a result, in order that $d\xi/d\varepsilon$ and $d\eta/d\varepsilon$ should exist, the rank of the matrix

$$\left\| \begin{matrix} F_\xi & F_\eta & F_\varepsilon \\ G_\xi & G_\eta & G_\varepsilon \end{matrix} \right\|$$

must be 1. But this is equivalent to the requirement that

(3.8)
$$\int_0^L \frac{q(s)}{W(s)} f(s, x_0(s), x_0'(s), 0)\,ds = 0.$$

Assuming (3.8) to hold, the table of first partial derivatives becomes

(3.9)
$$F_\xi = e^{-\alpha L} - 1, \quad F_\eta = 0, \quad F_\varepsilon = \frac{e^{-\alpha L} B_1}{g(A, 0)},$$
$$G_\xi = g(A, 0)\,q'(0)\,(e^{-\alpha L} - 1), \quad G_\eta = 0, \quad G_\varepsilon = e^{-\alpha L} q'(0)\,B_1$$

where we have let B_1 stand for the integral

$$\int_0^L \frac{p(s)}{W(s)} f(s, x_0(s), x_0'(s), 0)\, ds.$$

We can deduce from (3.9) that at $\varepsilon = 0$

$$\frac{d\xi}{d\varepsilon} = \frac{1}{e^{\alpha L} - 1} \frac{B_1}{g(A, 0)}. \tag{3.10}$$

To prove that $F = G = 0$ actually can be solved for ξ and η in terms of ε, and to evaluate the derivative $d\eta/d\varepsilon$ at $\varepsilon = 0$, we require the second partial derivatives of F and G at $(0, 0, 0)$.

As in Section 2.3 we solve the equation $F(\xi, \eta, \varepsilon) = 0$ for ξ as a function of η and ε. Denoting the result by $\xi = H(\eta, \varepsilon)$, we define the function $J(\eta, \varepsilon)$ by

$$J(\eta, \varepsilon) = G(H(\eta, \varepsilon), \eta, \varepsilon).$$

We then show that $J(\eta, \varepsilon) = 0$ can be solved for η as a function of ε for small ε. Then using $\xi = H(\eta, \varepsilon)$ we have ξ determined as a function of ε for small ε. This will effect the proof of existence of the periodic solution sought.

Before giving the values of the second partial derivatives of F and G at $(0, 0, 0)$, it will be convenient to discuss the term $x_1(t)$ in (3.1). We find that $x_1(t)$ must be a periodic solution of

$$L(y) = f(t, x_0(t), x_0'(t), 0) \tag{3.11}$$

where $L(y)$ is defined in (3.2). A necessary and sufficient condition that (3.11) have a solution of period L is that $f(t, x_0(t), x_0'(t), 0)$ be orthogonal on $(0, L)$ to the periodic solution of the homogeneous equation adjoint to (3.2). But this solution is exactly $q(t)/W(t)$, so that the orthogonality condition is exactly (3.8). Assuming the condition to be fulfilled, the most general periodic solution of (3.11) is found to be

$$\begin{aligned} y = Cp(t) &+ \frac{B_1}{(e^{\alpha L} - 1) g(A, 0)} \phi(t) \\ &+ \int_0^t \frac{p(s)q(t) - q(s)p(t)}{W(s)} f(s, x_0(s), x_0'(s), 0)\, ds. \end{aligned} \tag{3.12}$$

In (3.12), C is an arbitrary constant; $\phi(t)$ is defined in (3.7). The integral in (3.12) is exactly the partial derivative $x_\varepsilon(t, 0, 0, 0)$, and we shall use this notation later. When $C = 0$, we obtain the unique periodic solution of (3.11) which has $y'(0) = 0$. We denote this solution by $h_1(t)$, so that (3.12) can be written

(3.12) $$y = Cp(t) + h_1(t).$$

In the computation of the second partial derivatives of $J(\eta, \varepsilon)$ at $(0, 0)$, it is found that the second partial derivatives of F and G always occur in the combination

$$G_{\alpha\beta}(0, 0, 0) - g(A, 0)\,q'(0)\,F_{\alpha\beta}(0, 0, 0).$$

This combination turns out to be easier to evaluate than the partial derivatives $G_{\alpha\beta}$ and $F_{\alpha\beta}$ separately, so we give only the value of the combination.

$$G_{\xi\xi} - g(A, 0)\,q'(0)\,F_{\xi\xi} = g(A, 0)\int_0^L \frac{q(s)}{W(s)}\{-g_{xx}(x_0(s), x_0'(s))\,\phi(s)^2$$
$$- 2g_{xx'}(x_0(s), x_0'(s))\,\phi(s)\phi'(s) - g_{x'x'}(x_0(s), x_0'(s))\,\phi'(s)^2\}ds,$$

$$G_{\xi\eta} - g(A, 0)\,q'(0)\,F_{\xi\eta} = \frac{q'(L)\phi'(L) - q(L)\phi''(L)}{W(L)} - \frac{q'(0)\phi'(0) - q(0)\phi''(0)}{W(0)}$$

$$G_{\eta\eta} - g(A, 0)\,q'(0)\,F_{\eta\eta} = 0,$$

$$G_{\xi\varepsilon} - g(A, 0)\,q'(0)\,F_{\xi\varepsilon} = g(A, 0)\int_0^L \frac{q(s)}{W(s)}\{-g_{xx}(x_0(s), x_0'(s))\,\phi(s)x_\varepsilon(s)$$
$$- g_{xx'}(x_0(s), x_0'(s))\,(\phi(s)x_\varepsilon'(s) + \phi'(s)x_\varepsilon(s))$$

(3.14) $$- g_{x'x'}(x_0(s), x_0'(s))\,\phi'(s)x_\varepsilon'(s) + f_x(s, x_0(s), x_0'(s), 0)\,\phi(s)$$
$$+ f_{x'}(s, x_0(s), x_0'(s), 0)\,\phi'(s)\}ds,$$

$$G_{\eta\varepsilon} - g(A, 0)\,q'(0)\,F_{\eta\varepsilon} = \int_0^L \frac{q(s)}{W(s)} f_t(s, x_0(s), x_0'(s), 0)\,ds$$
$$+ \frac{q'(L)x_\varepsilon'(L) - q(L)x_\varepsilon''(L)}{W(L)} - \frac{q'(0)x_\varepsilon'(0) - q(0)x_\varepsilon''(0)}{W(0)}$$

$$G_{\varepsilon\varepsilon} - g(A,0)q'(0)F_{\varepsilon\varepsilon} = g(A,0)\int_0^L \frac{q(s)}{W(s)} \{-g_{xx}(x_0(s),x_0'(s))x_\varepsilon(s)^2$$

(3.14) (continued)
$$- 2g_{xx'}(x_0(s),x_0'(s))x_\varepsilon(s)x'_\varepsilon(s) - g_{x'x'}(x_0(s),x_0'(s))x'_\varepsilon(s)^2$$

$$+ 2f_x(s,x_0(s),x_0'(s),0)x_\varepsilon(s) + 2f_{x'}(s,x_0(s),x_0'(s),0)x'_\varepsilon(s)$$

$$+ 2f_\varepsilon(s,x_0(s),x_0'(s),0)\}ds.$$

Using the results of (3.14) we find the various partial derivatives of $J(\eta,\varepsilon)$ at $(0,0)$ to be

$$J_\eta(0,0) = 0, \qquad J_\varepsilon(0,0) = 0$$

$$J_{\eta\eta}(0,0) = 0$$

$$J_{\eta\varepsilon}(0,0) = \int_0^L \frac{q(s)}{W(s)} f_t(s,x_0(s),x_0'(s),0)\,ds,$$

$$J_{\varepsilon\varepsilon}(0,0) = g(A,0)\int_0^L \frac{q(s)}{W(s)} \{-g_{xx}(x_0(s),x_0'(s))h_1(s)^2$$

(3.15)
$$- 2g_{xx'}(x_0(s),x_0'(s))h_1(s)h_1'(s) - g_{x'x'}(x_0(s),x_0'(s))h_1'(s)^2$$

$$+ 2f_x(s,x_0(s),x_0'(s),0)h_1(s) + 2f_{x'}(s,x_0(s),x_0'(s),0)h_1'(s)$$

$$+ 2f_\varepsilon(s,x_0(s),x_0'(s),0)\}ds.$$

Hence near $\eta = \varepsilon = 0$, we have

$$J(\eta,\varepsilon) = J_{\eta\varepsilon}\eta\varepsilon + \frac{1}{2}J_{\varepsilon\varepsilon}\varepsilon^2 + \text{higher order terms.}$$

If the coefficient $J_{\eta\varepsilon}$ is not zero, we may solve $J(\eta,\varepsilon) = 0$ for η as a function of ε with

(3.16)
$$\eta = 0, \qquad \frac{d\eta}{d\varepsilon} = -\frac{J_{\varepsilon\varepsilon}}{2J_{\eta\varepsilon}}, \qquad (\varepsilon = 0)$$

by a modification of the implicit function theorem ([9] Lemma 8).

Thus when $J_{\eta\varepsilon} \neq 0$ we have shown that the equations $F = G = 0$ can be solved for ξ and η as functions of ε near $\varepsilon = 0$ with the derivatives $d\xi/d\varepsilon$ and $d\eta/d\varepsilon$ at $\varepsilon = 0$ given by (3.10) and (3.16) respectively.

As we saw in Part II, the initial values of $x_1(t)$ and $x_1'(t)$ are given by $d\xi/d\varepsilon$ and $d\eta/d\varepsilon$ at $\varepsilon = 0$ respectively. We can also determine these same initial values by formal manipulations with the perturbation series as in Part II. Since $x_1(t)$ is a periodic solution of (3.11), we have

$$x_1(t) = C_1 p(t) + h_1(t) \tag{3.17}$$

for some constant C_1. We can check immediately that $x_1(0) = h_1(0)$ has the value given by (3.10) by comparison with (3.12).

Now the value of C_1 is determined from the orthogonality condition for the existence of the term $x_2(t)$ in the perturbation series. The equation for $x_2(t)$ is

$$\begin{aligned} L(x_2) = & -\frac{1}{2} g_{xx}(x_0(t), x_0'(t))x_1(t)^2 - g_{xx'}(x_0(t), x_0'(t))x_1(t)x_1'(t) \\ & - \frac{1}{2} g_{x'x'}(x_0(t), x_0'(t))x_1'(t)^2 + f_x(t, x_0(t), x_0'(t), 0)x_1(t) \\ & + f_{x'}(t, x_0(t), x_0'(t), 0)x_1'(t) + f_\varepsilon(t, x_0(t), x_0'(t), 0). \end{aligned} \tag{3.18}$$

When (3.17) is substituted in the right member of (3.18), let the right member be called $R(t, C_1)$. Then the orthogonality condition for the existence of a periodic $x_2(t)$ is

$$\int_0^L \frac{q(s)}{W(s)} R(s, C_1)\, ds = 0,$$

which is an equation for C_1. On expansion and simplification it is found to reduce to

$$-C_1 J_{\eta\varepsilon} + \frac{J_{\varepsilon\varepsilon}}{2g(A, 0)} = 0$$

which gives for C_1 the value

$$C_1 = \frac{J_{\varepsilon\varepsilon}}{2J_{\eta\varepsilon}\, g(A, 0)} \tag{3.19}$$

This checks with the earlier value for $x_1'(0)$. From (3.17), $x_1'(0) = C_1 p'(0) = -g(A, 0)C_1$, and on comparison (3.19) and (3.16) are seen to agree.

THEOREM 3.1. *In equation* (1.2) *let* $g(x, x') \in C^2$ *and* $f(t, x, x', \varepsilon) \in C^1$. *Let equation* (1.1) *have the nonconstant periodic solution* $x_0(t)$ *of least period* L_0. *Let* $f(t, x, x', \varepsilon)$ *have least period* T *in* t *where* T *is a rational multiple of* L_0. *Let* L *be the least common multiple of* T *and* L_0. *Let* $x_0(0) = A, x_0'(0) = 0$. *Further let*

$$\alpha = \frac{1}{L}\int_0^L g_{x'}(x_0(t), x_0'(t)\, dt \neq 0, \tag{3.20}$$

$$\int_0^L \frac{q(s)}{W(s)} f(s, x_0(s), x_0'(s), 0)\, ds = 0, \tag{3.21}$$

$$\int_0^L \frac{q(s)}{W(s)} f_t(s, x_0(s), x_0'(s), 0)\, ds \neq 0. \tag{3.22}$$

Then (1.2) *has a unique periodic solution of period* L *of the form*

$$x = x_0(t) + \varepsilon x_1(t) + o(\varepsilon) \tag{3.23}$$

where $x_1(t)$ *is a uniquely determined solution of equation* (3.11) *namely*

$$x_1(t) = \frac{J_{\varepsilon\varepsilon}\, p(t)}{2J_{\eta\varepsilon}\, g(A, 0)} + h_1(t). \tag{3.24}$$

Proof: The proof of this theorem is contained in the preceding discussion.

Remarks: Condition (3.22) is equivalent to the nonvanishing of the critical Jacobian in the theory of Coddington and Levinson ([3] Chap. 14) and Friedrichs [5].

In case (3.21) is not satisfied, it may be that a translation of $x_0(t)$ will serve. This will be investigated in a later part of this section.

In case (3.22) is not satisfied, we can also obtain information in some cases using higher derivatives of F and G at (0, 0, 0). This will be done later in the section.

Before proceeding to higher derivatives of F and G at (0, 0, 0) it is necessary to discuss the function $x_2(t)$ and equation (3.19), the condition for existence of $x_2(t)$. If condition (3.22) in Theorem 3.1 is not satisfied, no value for C_1 can be found from (3.19). Indeed if $J_{\eta\varepsilon} = 0$ and $J_{\varepsilon\varepsilon} \neq 0$, the orthogonality condition for the existence of $x_2(t)$ cannot be satisfied, no value of C_1 exists, and a periodic solution of the type sought does not exist. However if $J_{\eta\varepsilon} = J_{\varepsilon\varepsilon} = 0$, the orthogonality condition is satisfied identically, we can find $x_2(t)$, and there is a possibility of obtaining the desired periodic solutions of (1.2).

Let $\phi_2(t)$ be that periodic solution of

$$(3.25)\qquad L(y) = -g_{xx}(x_0(t), x_0'(t))\,p(t)^2 - 2g_{xx'}(x_0(t), x_0'(t))\,p(t)\,p'(t) - g_{x'x'}(x_0(t), x_0'(t))\,p'(t)^2$$

which satisfies $y'(0) = 0$.

Let $k_2(t)$ be that periodic solution of

$$(3.26)\qquad \begin{aligned} L(y) = {} & -g_{xx}(x_0(t), x_0'(t))\,p(t)\,h_1(t) - g_{xx'}(x_0(t), x_0'(t)[p(t)\,h_1'(t) + p'(t)h_1(t)] \\ & - g_{x'x'}(x_0(t), x_0'(t))\,p'(t)\,h_1'(t) + f_x(t, x_0(t), x_0'(t), 0)\,p(t) \\ & + f_{x'}(t, x_0(t), x_0'(t), 0)\,p'(t) \end{aligned}$$

satisfying $y'(0) = 0$. Finally let $h_2(t)$ be that periodic solution of

$$(3.27)\qquad \begin{aligned} L(y) = {} & -\frac{1}{2}g_{xx}(x_0(t), x_0'(t))\,h_1(t)^2 - g_{xx'}(x_0(t), x_0'(t))\,h_1(t)\,h_1'(t) \\ & -\frac{1}{2}g_{x'x'}(x_0(t), x_0'(t))\,h_1'(t)^2 + f_x(t, x_0(t), x_0'(t), 0)\,h_1(t) \\ & + f_{x'}(t, x_0(t), x_0'(t), 0), 0)\,h_1'(t) + f_\varepsilon(t, x_0(t), x_0'(t), 0) \end{aligned}$$

satisfying $y'(0) = 0$. On comparison with (3.17) and (3.18), we see that a periodic solution of (3.18) must have the form

$$(3.28)\qquad x_2(t) = C_2\,p(t) + \frac{1}{2}C_1^2\,\phi_2(t) + C_1\,k_2(t) + h_2(t).$$

In the cáse of interest to us, when $J_{\eta\varepsilon} = J_{\varepsilon\varepsilon} = 0$, the periodic solutions $\phi_2(t)$, $k_2(t)$, and $h_2(t)$ all exist; indeed the orthogonality conditions are $J_{\eta\eta} = 0$, $J_{\eta\varepsilon} = 0$, and $J_{\varepsilon\varepsilon} = 0$ respectively. In (3.28) the coefficient C_2 is an as yet undetermined constant.

The function $\phi_2(t)$ can be explicitly written. A periodic solution of (3.25) is $y = p'(t)$. To obtain $y'(0) = 0$, we add an appropriate multiple of $p(t)$. It is found that

$$\phi_2(t) = p'(t) + g_{x'}(A, 0)\,p(t).$$

When all the second derivatives of $J(\eta, \varepsilon)$ at $(0, 0)$ are zero, it is necessary to use the third derivatives of J to solve $J(\eta, \varepsilon) = 0$ for η as a function of ε. When all the first and second derivatives of J are zero, the third derivatives of J at $(0, 0)$ have the values:

$$J_{\eta\eta\eta} = 0,$$

$$J_{\eta\eta\varepsilon} = \frac{1}{g(A,0)} \int_0^L \frac{q(s)}{W(s)} f_{tt}(s, x_0(s), x_0'(s), 0)\, ds,$$

$$J_{\eta\varepsilon\varepsilon} = 2\int_0^L \frac{q(s)}{W(s)} \{- g_{xx}(x_0(s), x_0'(s)) h_1(s) h_1'(s) - k_2(s))$$
$$- g_{xx'}(x_0(s), x_0'(s))[h_1(s) h_1''(s) - k_2'(s)) + h_1'(s)$$
$$(h_1'(s) - k_2(s))] - g_{x'x'}(x_0(s), x_0'(s)) h_1'(s)$$
$$(h_1''(s) - k_2'(s)) + f_x(s, x_0(s), x_0'(s), 0)(h_1'(s) - k_2(s))$$
$$+ f_{x'}(s, x_0(s), x_0'(s), 0)(h_1''(s) - k_2'(s))$$
$$+ f_{tx}(s, x_0(s), x_0'(s), 0) h_1(s) + f_{tx'}(s, x_0(s), x_0'(s), 0) h_1'(s)$$
$$+ f_{t\varepsilon}(s, x_0(s), x_0'(s), 0)\} ds,$$

$$J_{\varepsilon\varepsilon\varepsilon} = 6g(A,0) \int_0^L \frac{q(s)}{W(s)} \{- g_{xx}(x_0(s), x_0'(s)) h_1(s) h_2(s)$$

(3.29) $$g_{xx'}(x_0(s), x_0'(s))[h_1(s) h_2'(s) + h_1'(s) h_2(s)$$

$$- g_{x'x'}(x_0(s), x_0'(s)) h_1'(s) h_2'(s) - \frac{1}{6} g_{xxx}(x_0(s), x_0'(s)) h_1(s)^3$$
$$- \frac{1}{2} g_{xxx'}(x_0(s), x_0'(s)) h_1(s)^2 h_1'(s)$$
$$- \frac{1}{2} g_{xx'x'}(x_0(s), x_0'(s)) h_1(s) h_1'(s)^2$$
$$- \frac{1}{6} g_{x'x'x'}(x_0(s), x_0'(s)) h_1'(s)^3 + f_x(s, x_0(s), x_0'(s), 0) h_2(s)$$
$$+ f_{x'}(s, x_0(s), x_0'(s), 0) h_2'(s) + \frac{1}{2} f_{xx}(s, x_0(s), x_0'(s), 0) h_1(s)^2$$
$$+ f_{xx'}(s, x_0(s), x_0'(s), 0) h_1(s) h_1'(s)$$
$$+ \frac{1}{2} f_{x'x'}(s, x_0(s), x_0'(s), 0) h_2'(s)^2 + f_{x\varepsilon}(s, x_0(s), x_0'(s), 0) h_1(s)$$
$$+ f_{x'\varepsilon}(s, x_0(s), x_0'(s), 0) h_1'(s) + \frac{1}{2} f_{\varepsilon\varepsilon}(s, x_0(s), x_0'(s), 0)\} ds.$$

Since these third derivatives of J involve third derivatives of g and second derivatives of f, we assume for this part of the theory that $g \in C^3$ and $f \in C^2$.

If it is assumed that all of the second derivatives of J vanish at $(0, 0)$, we have that near $(0, 0)$

$$J(\eta, \varepsilon) = \frac{1}{2} J_{\eta\eta\varepsilon} \eta^2 \varepsilon + \frac{1}{2} J_{\eta\varepsilon\varepsilon} \eta\varepsilon^2 + \frac{1}{6} J_{\varepsilon\varepsilon\varepsilon} \varepsilon^3 + \text{higher order terms.} \tag{3.30}$$

Now suppose that the discriminant of the quadratic equation

$$\frac{1}{2} J_{\eta\eta\varepsilon} m^2 + \frac{1}{2} J_{\eta\varepsilon\varepsilon} m + \frac{1}{6} J_{\varepsilon\varepsilon\varepsilon} = 0 \tag{3.31}$$

is positive, so that (3.31) has one or two simple real roots. Then by Lemma 8 of [9] we can solve $J(\eta, \varepsilon) = 0$ for η as a function of ε for ε near zero, with $\eta = 0$ and $d\eta/d\varepsilon = m$, a root of equation (3.31). If (3.31) has two real simple roots, then we can find two different functions of ε which can be used for η.

We have again effected the solution of the system $F = G = 0$ for ξ and η as functions of ε, with the derivatives $d\xi/d\varepsilon$ and $d\eta/d\varepsilon$ at $\varepsilon = 0$ found from (3.10) and the roots of (3.31) respectively. We thus have proved the following theorem.

THEOREM 3.2. *In equation* (1.2) *let* $g(x, x') \in C^3$ *and* $f(t, x, x', \varepsilon) \in C^2$. *Let all hypotheses of Theorem* 3.1 *be satisfied except for* (3.22). *Let*

$$\int_0^L \frac{q(s)}{W(s)} f(s, x_0(s), x_0'(s), 0)\, ds = 0, \tag{3.32}$$

$$\int_0^L \frac{q(s)}{W(s)} f_t(s, x_0(s), x_0'(s), 0)\, ds = 0, \tag{3.33}$$

$$J_{\varepsilon\varepsilon}(0, 0) = 0, \tag{3.34}$$

$$J_{\eta\varepsilon\varepsilon}(0, 0)^2 - \frac{4}{3} J_{\eta\eta\varepsilon}(0, 0) J_{\varepsilon\varepsilon\varepsilon}(0, 0) > 0. \tag{3.35}$$

Then (1.2) *has one or two periodic solutions of period* L *of the form*

$$x = x_0(t) + \varepsilon x_1(t) + o(\varepsilon) \tag{3.36}$$

where $x_1(t)$ *is a uniquely determined solution of equation* (3.11), *according as*

$$\int_0^L \frac{q(s)}{W(s)} f_{tt}(s, x_0(s), x_0'(s), 0)\, ds \tag{3.37}$$

does or does not vanish.

As before we may also determine the derivatives $d\xi/d\varepsilon$ and $d\eta/d\varepsilon$ at $\varepsilon = 0$ by manipulation of the perturbation series. Under the hypotheses of Theorem 3.2 the orthogonality condition for the existence of the term $x_2(t)$ of the perturbation series is identically satisfied, and $x_2(t)$ is given by (3.2), where in (3.2) both C_1 and C_2 are unknown. We can determine C_1 by the orthogonality condition for the existence of the term $x_3(t)$ in the perturbation series. We omit the lengthy details, and merely state that the orthogonality condition reduces to

$$\frac{1}{2}g(A,0)\,J_{\eta\eta\varepsilon}C_1^2 - \frac{1}{2}J_{\eta\varepsilon\varepsilon}C_1 + \frac{1}{6g(A,0)}J_{\varepsilon\varepsilon\varepsilon} = 0, \tag{3.38}$$

which is equivalent to (3.31) when we recall that $x_1'(0)$ is equal to $d\eta/d\varepsilon$ at $\varepsilon = 0$ and also to $-g(A,0)\,C_1$.

The analysis of stability will be similar to that in Part II. If $x(t)$ is the periodic solution found in Theorems 3.1 or 3.2, the related variation equation is

$$y'' + g_{x'}(x,x')y' + g_x(x,x')y = \varepsilon f_x(t,x,x',\varepsilon)y + \varepsilon f_{x'}(t,x,x',\varepsilon)y' \tag{3.39}$$

The stability of the periodic solution $x(t)$ will be described by the characteristic multipliers of (3.39). See Section 1.4 for definitions.

Let $\phi(t)$ and $\psi(t)$ represent the solutions of (3.39) which have $\phi(0) = \psi'(0) = 1$, $\phi'(0) = \psi(0) = 0$. When $\varepsilon = 0$, (3.39) reduces to (3.2), and as we saw earlier, when $\varepsilon = 0$, $\phi(t) = g(A,0)\,q(t) + q'(0)\,p(t)$, and $\psi(t) = -p(t)/g(A,0)$, for Case II.

If we write $x(t)$ as $x_0(t) + \varepsilon x_1(t) + o(\varepsilon)$, equation (3.39) assumes the form

$$\begin{aligned} &y'' + g_{x'}(x_0(t),x_0'(t))y' + g_x(x_0(t),x_0'(t))y \\ &= \varepsilon\{-g_{xx}(x_0(t),x_0'(t))x_1(t)y - g_{xx'}(x_0(t),x_0'(t)) \\ &\quad [x_1'(t)y + x_1(t)y'] - g_{x'x'}(x_0(t),x_0'(t))x_1'(t)y' \\ &\quad + f_x(t,x_0(t),x_0'(t),0)y + f_{x'}(t,x_0(t),x_0'(t),0)y'\} \\ &\quad + o(\varepsilon). \end{aligned} \tag{3.40}$$

Now transform (3.40) to an integral equation for $\phi(t)$ or $\psi(t)$ and carry through the first step of successive approximations. If this is done for both $\phi(t)$ and $\psi(t)$, we find that

$$\phi(L) = e^{-\alpha L} + \varepsilon d_{11} + o(\varepsilon),$$

$$\phi'(L) = g(A, 0)q'(0)(e^{-\alpha L} - 1) + \varepsilon d_{21} + o(\varepsilon),$$

$$\psi(L) = -\varepsilon \frac{e^{-\alpha L}}{g(A, 0)^2} \int_0^L \frac{p(s)}{W(s)} \{-g_{xx}(x_0(s), x_0'(s))x_1(s)p(s)$$

$$- g_{xx'}(x_0(s), x_0'(s))[x_1'(s)p(s) + x_1(s)p'(s)]$$

$$- g_{x'x'}(x_0(s), x_0'(s))x_1'(s)p'(s) + f_x(s, x_0(s), x_0'(s), 0)p(s)$$

$$+ f_{x'}(s, x_0(s), x_0'(s), 0)p'(s)\}ds + o(\varepsilon)$$

$$= \varepsilon d_{12} + o(\varepsilon), \tag{3.41}$$

$$\psi'(L) = 1 - \varepsilon \int_0^L \left[\frac{q'(0)e^{-\alpha L}}{g(A, 0)} \frac{p(s)}{W(s)} + \frac{q(s)}{W(s)}\right] \{-g_{xx}(x_0(s), x_0'(s))x_1(s)p(s)$$

$$- g_{xx'}(x_0(s), x_0'(s))[x_1'(s)p(s) + x_1(s)p'(s)]$$

$$- g_{x'x'}(x_0(s), x_0'(s))x'(s)p'(s) + f_x(s, x_0(s), x_0'(s), 0)p(s)$$

$$+ f_{x'}(s, x_0(s), x_0'(s), 0)p'(s)\}ds + o(\varepsilon)$$

$$= 1 + \varepsilon d_{22} + o(\varepsilon).$$

The values of d_{11} and d_{21} are of no significance in the first stability theorem.

Now the characteristic multipliers of (3.39) are the eigenvalues of the matrix

$$\left\| \begin{matrix} \phi(L) & \psi(L) \\ \phi'(L) & \psi'(L) \end{matrix} \right\|$$

One of them is $e^{-\alpha L}$ when $\varepsilon = 0$, so for small nonzero ε this characteristic multiplier remains near $e^{-\alpha L}$, and so its absolute value remains away from 1. The second characteristic multiplier is equal to 1 for $\varepsilon = 0$, and we must determine its behavior for small nonzero ε

Let

$$G(\varepsilon,\lambda) = \det \begin{Vmatrix} \phi(L) - \lambda & \psi(L) \\ \phi'(L) & \psi'(L) - \lambda \end{Vmatrix}$$

$$= \det \begin{Vmatrix} e^{-\alpha L} - \lambda + \varepsilon d_{11} + o(\varepsilon) & \varepsilon d_{12} + o(\varepsilon) \\ g(A,0)\, q'(0)(e^{-\alpha L} - 1) + \varepsilon d_{21} + o(\varepsilon) & 1 - \lambda + \varepsilon d_{22} + o(\varepsilon) \end{Vmatrix}$$

$G(0,1) = 0$. Hence if $G_\lambda(0,1) \neq 0$, we may solve the equation $G(\varepsilon,\lambda) = 0$ for λ as a function of ε, and so determine the behavior for ε near zero of that characteristic multiplier which reduces for $\varepsilon = 0$ to $+1$. We find

$$G_\lambda(0,1) = -(e^{-\alpha L} - 1) \neq 0,$$

$$G_\varepsilon(0,1) = d_{22}(e^{-\alpha L} - 1) - d_{12}\, g(A,0)\, q'(0)\,(e^{-\alpha L} - 1)$$

Therefore
$$\left(\frac{d\lambda}{d\varepsilon}\right)_{\varepsilon=0} = -\frac{G_\varepsilon(0,1)}{G_\lambda(0,1)} = d_{22} - d_{12}\, g(A,0)\, q'(0)$$

$$= -\int_0^L \frac{q(s)}{W(s)} \{ - g_{xx}(x_0(s), x_0'(s))\, x_1(s)\, p(s)$$
$$- g_{xx'}(x_0(s), x_0'(s))\,[x_1'(s)\, p(s) + x_1(s)\, p'(s)]$$
$$- g_{x'x'}(x_0(s), x_0'(s))\, x_1'(s)\, p'(s) + f_x(s, x_0(s), x_0'(s), 0)\, p(s)$$
$$+ f_{x'}(s, x_0(s), x_0'(s), 0)\, p'(s) \}\, ds.$$

This last integral simplifies, and the final result is

$$\left(\frac{d\lambda}{d\varepsilon}\right)_{\varepsilon=0} = \int_0^L \frac{q(s)}{W(s)}\, f_t(s, x_0(s), x_0'(s), 0)\, ds.$$

Thus for small ε, the characteristic multiplier has the form

$$\lambda = 1 + \varepsilon \int_0^L \frac{q(s)}{W(s)}\, f_t(s, x_0(s), x_0'(s), 0)\, ds + o(\varepsilon). \tag{3.42}$$

THEOREM 3.3. *Let the hypotheses of Theorem* 3.1 *be satisfied. Then the periodic solution found in this theorem has the following stability properties.*

(a) *If* $\alpha < 0$, *and*

$$(3.43) \qquad \varepsilon \int_0^L \frac{q(s)}{W(s)} f_t(s, x_0(s), x_0'(s), 0)\,ds$$

is positive, the solution is completely unstable.

(b) *If* $\alpha < 0$ *and* (3.43) *is negative, or if* $\alpha > 0$ *and* (3.43) *is positive, the solution is directly unstable.*

(c) *If* $\alpha > 0$, *and* (3.43) *is negative, the solution is completely stable.*

Proof of Theorem 3.3: The characteristic multiplier near to $e^{-\alpha L}$ has absolute value greater or less than 1 according as α is negative or positive. The second characteristic multiplier has absolute value greater or less than 1 for small nonzero ε according as (3.43) is positive or negative by (3.42). Hence the various combinations of signs give the stabilities as asserted. This completes the proof.

We now determine the stability of the periodic solutions found in Theorem 3.2. In this case, (3.43) is zero, so that the first derivative $\frac{d\lambda}{d\varepsilon}$ of the characteristic multiplier vanishes at $\varepsilon = 0$. To determine the behavior of the characteristic multiplier near $\varepsilon = 0$, we need the second derivative of λ.

To do this, we need to carry the successive approximation solution of the variation equations for $\phi(t)$ and $\psi(t)$ a second step. If this is done we obtain

$$\phi(L) = e^{-\alpha L} + \varepsilon d_{11} + \varepsilon^2 f_{11} + o(\varepsilon^2),$$

$$\phi'(L) = g(A, 0)\,q'(0)\,(e^{-\alpha L} - 1) + \varepsilon d_{21} + \varepsilon^2 f_{21} + o(\varepsilon^2),$$

$$\psi(L) = \varepsilon d_{12} + \varepsilon^2 f_{12} + o(\varepsilon^2)$$

$$\psi'(L) = 1 + \varepsilon d_{22} + \varepsilon^2 f_{22} + o(\varepsilon^2),$$

where the d_{ij} are as in (3.41), and the f_{ij} are more complicated expressions involving third derivatives of g and second derivatives of f. Using the notation introduced earlier, and using the fact that $G_\varepsilon(0, 1) = 0$ in our present case, we find that

$$(3.44) \qquad \left(\frac{d^2\lambda}{d\varepsilon^2}\right)_{\varepsilon=0} = -\frac{G_{\varepsilon\varepsilon}(0,1)}{G_\lambda(0,1)} = 2(f_{22} - g(A,0)\,q'(0)\,f_{12}) + \frac{2d_{12}}{e^{-\alpha L} - 1}\,(g(A,0)\,q'(0)\,d_{11} - d_{21})$$

Equation (3.44) can be much simplified, although the computation is long and tedious. The final result is

$$(3.45) \qquad \left(\frac{d^2\lambda}{d\varepsilon^2}\right)_{\varepsilon=0} = -2C_1 g(A, 0) J_{\eta\eta\varepsilon} + J_{\eta\varepsilon\varepsilon}.$$

We note that the right member of (3.45) is the derivative with respect to C_1 of the left member of equation (3.38) apart from a factor of $-\frac{1}{2}$, and so does not vanish under the hypotheses of Theorem 3.2, since C_1 is a simple root of equation (3.38).

THEOREM 3.4. *Let all of the hypotheses of Theorem* 3.2 *be satisfied, Then a periodic solution found under these conditions has the following stability properties:*

(a) *If* $\alpha < 0$, *and if*

$$(3.46) \qquad -2C_1 g(A, 0) J_{\eta\eta\varepsilon} + J_{\eta\varepsilon\varepsilon}$$

is positive, the solution is completely unstable.

(b) *If* $\alpha > 0$ *and* (3.46) *is positive, or if* $\alpha < 0$ *and* (3.46) *is negative, the solution is directly unstable.*

(c) *If* $\alpha > 0$ *and* (3.46) *is negative, the solution is completely stable.*

Proof of Theorem 3.4. Since at $\varepsilon = 0$, $\frac{d\lambda}{d\varepsilon} = 0$, and $\frac{d^2\lambda}{d\varepsilon^2}$ is given by (3.46), the characteristic multiplier which reduces to 1 for $\varepsilon = 0$ is greater or less than 1 for small ε according as (3.46) is positive or negative. The second characteristic multiplier is near to $e^{-\alpha T}$ and so is greater or less than 1 according as α is negative or positive. This completes the proof.

Remark. In case Theorem 3.2 gives two periodic solutions for small ε, the stability will be different for the two solutions. The expression (3.46) will have opposite signs for the two different values of C_1.

We now turn to the case of a periodic solution which reduces not to $x_0(t)$, but to some translation $x_0(t+\tau)$. Because our theory has been built around $x_0(t)$, it is easier to consider the problem

$$(3.47) \qquad x'' + g(x, x') = \varepsilon f(t-\tau, x, x', \varepsilon).$$

A solution of (3.47) near to $x_0(t)$ is the same as a solution of (1.2) near to $x_0(t+\tau)$.

Let $F_1(\tau)$ be defined by

$$F_1(\tau) = \int_0^L \frac{q(t)}{W(t)} f(t-\tau, x_0(t), x_0'(t), 0)\,dt. \tag{3.48}$$

We note that

$$F_1'(\tau) = -\int_0^L \frac{q(t)}{W(t)} f_t(t-\tau, x_0(t), x_0'(t), 0)\,dt, \tag{3.49}$$

and that

$$F_1''(\tau) = \int_0^L \frac{q(t)}{W(t)} f_{tt}(t-\tau, x_0(t), x_0'(t), 0)\,dt. \tag{3.50}$$

Using $F_1(\tau)$ we can state a theorem equivalent to Theorems 3.1 and 3.3.

THEOREM 3.5. *Let all of the hypotheses of Theorem* 3.1 *preceding equation* (3.21) *hold. Let* $F_1(\tau) = 0$ *and* $F_1'(\tau) \neq 0$. *Then there exists for small non-zero* ε *a periodic solution of*

$$x'' + g(x, x') = \varepsilon f(t-\tau, x, x', \varepsilon) \tag{3.47}$$

of the form

$$x = x_0(t) + \varepsilon x_1(t) + o(\varepsilon), \tag{3.51}$$

and correspondingly a periodic solution of

$$x'' + g(x, x') = \varepsilon f(t, x, x', \varepsilon) \tag{3.52}$$

of the form

$$x = x_0(t+\tau) + \varepsilon x_1(t+\tau) + o(\varepsilon). \tag{3.53}$$

The characteristic multiplier $\lambda(\varepsilon)$ *of the variational equation which reduces to* $+1$ *for* $\varepsilon = 0$ *is greater or less than* $+1$ *for small nonzero* ε *according as* $\varepsilon F_1'(\tau)$ *is negative or positive. Accordingly the stability of the periodic solution is as in the following table.*

Sign of α	Sign of $\varepsilon F_1'(\tau)$	Stability
+	+	Completely stable
+	-	Directly unstable
-	+	Directly unstable
-	-	Completely unstable

Proof of Theorem 3.5. In view of (3.49), the present theorem is just a restatement of Theorems 3.1 and 3.3 with different notation.

Because $x_0(t)$ has period L_0, and because $f(t, x, x', \varepsilon)$ has period T in its first variable, $F_1(\tau)$ is a periodic function of τ of which both L_0 and T are periods. Hence if $pT = qL_0 = L$, where p and q are relatively prime integers, $F_1(\tau)$ will have the period $\frac{T}{q} = \frac{L_0}{p} = \frac{L}{pq}$.

Suppose that all of the zeros of $F_1(\tau)$ are simple. Then there will be an even number of them in the interval $0 \leqq \tau < \frac{L_0}{p}$, and thus an even multiple of p in the interval $0 \leqq \tau < L_0$. Therefore there will be a multiple of $2p$ translations of $x_0(t)$ near to which there exist periodic solutions of (3.52) in the form (3.53). Moreover since the sign of $F_1'(\tau)$ will alternate at successive simple zeros, half of the solutions found will be directly unstable. The other half will be completely stable or completely unstable according as α is positive or negative. Since the period of the periodic solutions is $L = pT$, they are subharmonics of order p.

We can also obtain an analogous theorem corresponding to Theorems 3.2 and 3.4. We still have the requirement that $F_1(\tau) = 0$. If τ is an isolated multiple zero of $F_1(\tau)$, so that Theorem 3.5 does not apply, but the conditions of Theorem 3.2 hold, we can regard the situation as a limiting case of Theorem 3.5 as two or more periodic solutions come into coalescence.

The theorem we obtain is for the case that $F_1(\tau) \equiv 0$. Assuming this, we have that the equation satisfied by $x_1(t)$

$$(3.54) \qquad L(y) = f(t - \tau, x_0(t), x_0'(t), 0)$$

will have a periodic solution for all values of τ. Let $h_1(t, \tau)$ be that periodic solution of (3.54) which satisfies the further condition $y'(0) = 0$. Let $F_2(\tau)$ be defined by

$$(3.55) \qquad F_2(\tau) = \int_0^L \frac{q(s)}{W(s)} \{- g_{xx}(x_0(s), x_0'(s)) h_1(s, \tau)^2$$

(3.55) (continued)
$$- 2g_{xx'}(x_0(s), x_0'(s))\, h_1(s, \tau)\, h_1'(s, \tau) - g_{x'x'}(x_0(s), x_0'(s))\, h_1'(s, \tau)^2$$

$$+ 2f_x(s-\tau, x_0(s), x_0'(s), 0)\, h_1(s, \tau) + 2f_{x'}(s-\tau, x_0(s)\, x_0'(s), 0)\, h_1'(s, \tau)$$

$$+ 2f_\varepsilon(s-\tau, x_0(s), x_0'(s), 0)\}\, ds.$$

Note that $F_2(\tau)$ is just $J_{\varepsilon\varepsilon}(0, 0)$ apart from a factor of $g(A, 0)$, where we have used $h_1(t, \tau)$ for $h_1(t)$ and $f(t-\tau, x, x', \varepsilon)$ for $f(t, x, x', \varepsilon)$. A prime on $h_1(t, \tau)$ denotes differentiation with respect to t. From the hypothesis of Theorem 3.2 we see that a requirement will be that $F_2(\tau) = 0$.

The discriminant condition of Theorem 3.2 can also be expressed in terms of $F_2(\tau)$. Because $F_1(\tau) \equiv 0$, it follows from (3.50) that $J_{\eta\eta\varepsilon}(0, 0) = 0$, so that the discriminant is positive if and only if $J_{\eta\varepsilon\varepsilon}(0, 0) \neq 0$. What is more, the quantity $J_{\eta\varepsilon\varepsilon}(0, 0)$ will be the critical stability quantity as seen from Theorem 3.4.

If (3.55) is differentiated with respect to τ, and appropriate simplifications are made, it is found that

(3.56)
$$F_2'(\tau) = - J_{\eta\varepsilon\varepsilon}(0, 0)$$

where in the formula for $J_{\eta\varepsilon\varepsilon}(0, 0)$ in (3.29) we replace $f(t, x, x', \varepsilon)$ by $f(t-\tau, x, x', \varepsilon)$, $h_1(t)$ by $h_1(t, \tau)$, and $k_2(t)$ by $k_2(t, \tau)$, where $k_2(t, \tau)$ is that periodic solution of

(3.57)
$$\begin{aligned} L(y) = &- g_{xx}(x_0(t), x_0'(t))\, p(t)\, h_1(t, \tau) \\ &- g_{xx'}(x_0(t), x_0'(t))\, [p(t)\, h_1'(t, \tau) + p'(t)\, h_1(t, \tau)] \\ &- g_{x'x'}(x_0(t), x_0'(t))\, p'(t)\, h_1'(t, \tau) + f_x(t-\tau, x_0(t), x_0'(t), 0)\, p(t) \\ &+ f_{x'}(t-\tau, x_0(t), x_0'(t), 0)\, p'(t), \end{aligned}$$

which satisfies $y'(0) = 0$. The orthogonality condition for the existence of a periodic solution of (3.57) is found to be exactly $F_1'(\tau) = 0$, so that $k_2(t, \tau)$ surely exists for all τ.

THEOREM 3.6. *Let all of the continuity hypotheses of Theorem* 3.2 *hold, and let all of the hypotheses of Theorem* 3.1 *preceding equation* (3.21) *hold: Let* $F_1(\tau) \equiv 0$. *Let* $F_2(\tau) = 0$, *and* $F_2'(\tau) \neq 0$. *Then there exists a periodic solution of*

(3.47) $$x'' + g(x, x') = \varepsilon f(t - \tau, x, x', \varepsilon)$$

of the form

(3.51) $$x = x_0(t) + \varepsilon x_1(t) + o(\varepsilon),$$

and correspondingly a periodic solution of

(3.52) $$x'' + g(x, x') = \varepsilon f(t, x, x', \varepsilon)$$

of the form

(3.53) $$x = x_0(t + \tau) + \varepsilon x_1(t + \tau) + o(\varepsilon).$$

The characteristic multiplier $\lambda(\varepsilon)$ *of the variational equation which reduces to* $+1$ *for* $\varepsilon = 0$ *is greater or less than* $+1$ *for small nonzero* ε *according as* $F_2'(\tau)$ *is negative or positive. Accordingly the stability of the periodic solution is as in the following table.*

Sign of α	Sign of $F_2'(\tau)$	Stability
+	+	Completely stable
+	-	Directly unstable
-	+	Directly unstable
-	-	Completely unstable

Proof of Theorem 3.6. This is essentially a restatement of Theorems 3.2 and 3.4 in new notation. The stability result uses (3.56) to relate $F_2'(\tau)$ to the hypotheses of Theorem 3.4

3.3. Existence and Stability in Case III. It was seen in Section 3.1 that Case III occurs when

$$g_{x'}(x_0(t), x_0'(t))\, dt = 0,$$

but the variation equation (3.2) does not have all its solutions periodic. The characteristic multipliers of (3.2) are both equal to +1.

The present section represents a generalization of some of the results of [8], in which $g(x, x')$ depended only on x, and $f(t, x, x', \varepsilon)$ depended only on t. In the appendix to this section we give the computations for some of the partial derivatives of F and G to illustrate the method.

We must again choose a fundamental pair of solutions of the variation equation. One of them will be $p(t) \equiv x_0'(t)$, as in Section 3.2. The relations

$$(3.4) \qquad p(0) = p(L) = 0$$
$$p'(0) = p'(L) = -g(A, 0)$$

are still true. By Floquet theory, a second solution of (3.2) exists in Case III of the form $tp(t) + r(t)$, where $r(t)$ is likewise periodic. For our second solution we choose $q(t)$, defined by the initial conditions $q(0) = 1/g(A, 0)$, and $q'(0) = 0$. There exists then a number $K \neq 0$ such that

$$(3.58) \qquad q(t) = Ktp(t) + r(t)$$

where $r(t)$ has period L. The fact that $K \neq 0$ is what distinguishes Case III from Case IV. If $K = 0$, then $q(t)$, and hence all solutions of (3.2), would have period L. We find that

$$(3.59) \qquad \begin{aligned} q(0) &= 1/g(A, 0) & q'(0) &= 0 \\ q(L) &= 1/g(A, 0) & q'(L) &= -KLg(A, 0). \end{aligned}$$

Because of the selection of initial values, the Wronskian

$$W(t) = e^{-\int_0^t g_{x'}(x_0(s), x_0'(s))\,ds}$$

In this case, $W(L) = 1$, so that $W(t)$ is periodic with period L (and least period L_0).

The solutions $\phi(t)$ and $\psi(t)$ of (3.2) are given by

$$(3.60) \qquad \phi(t) = g(A, 0)\, q(t), \qquad \psi(t) = -\frac{p(t)}{g(A, 0)}.$$

The first partial derivatives of F and G at $(0, 0, 0)$ have the values

$$(3.61) \qquad \begin{aligned} F_\xi &= 0, & F_\eta &= 0, \\ G_\xi &= -KLg(A, 0)^2, & G_\eta &= 0, \\ F_\varepsilon &= \frac{1}{g(A, 0)} \int_0^L \frac{p(s)}{W(s)} f(s, x_0(s), x_0'(s), 0)\,ds, \\ G_\varepsilon &= -KLg(A, 0) \int_0^L \frac{p(s)}{W(s)} f(s, x_0(s), x_0'(s), 0)\,ds \end{aligned}$$

$$+ g(A, 0) \int_0^L \frac{q(s)}{W(s)} f(s, x_0(s), x_0'(s), 0)\, ds.$$

The Jacobian of F and G with respect to ξ and η is zero, and the rank of the Jacobian matrix is 1. In order that the derivatives $d\xi/d\varepsilon$ and $d\eta/d\varepsilon$ exist, the matrix

$$\left\| \begin{matrix} F_\xi & F_\eta & F_\varepsilon \\ G_\xi & G_\eta & G_\varepsilon \end{matrix} \right\|$$

must also have rank 1. But this is equivalent to the requirement

$$(3.62) \qquad \int_0^L \frac{p(s)}{W(s)} f(s, x_0(s), x_0'(s), 0)\, ds = 0.$$

Assuming (3.62) to hold, the table of first partial derivatives becomes

$$(3.63) \qquad \begin{aligned} &F_\xi = 0, && F_\eta = 0, && F_\varepsilon = 0, \\ &G_\xi = -KLg(A, 0), && G_\eta = 0, && G_\varepsilon = g(A, 0) B_1 \end{aligned}$$

where we let B_1 stand for the integral

$$\int_0^L \frac{q(s)}{W(s)} f(s, x_0(s), x_0'(s), 0)\, ds.$$

We can deduce from (3.63) that at $\varepsilon = 0$,

$$(3.64) \qquad \frac{d\xi}{d\varepsilon} = \frac{B_1}{KLg(A, 0)}.$$

To effect the solution of the system $F = G = 0$ for ξ and η as functions of ε we proceed as earlier. Since $G_\xi \neq 0$, we solve $G(\xi, \eta, \varepsilon) = 0$ for ξ as a function of η and ε. Call the result $\xi = H(\eta, \varepsilon)$. Then form the function $J(\eta, \varepsilon)$ as

$$J(\eta, \varepsilon) = F(H(\eta, \varepsilon), \eta, \varepsilon).$$

We then solve the equation $J(\eta, \varepsilon) = 0$ for η as a function of ε, and obtain the derivative $d\eta/d\varepsilon$ at $\varepsilon = 0$. Once this is done, we have shown that $F = G = 0$ can indeed be solved for ξ and η as functions of ε near $\varepsilon = 0$, and we have proved the existence of the periodic solution sought.

Before discussing the second derivatives of F and G at $(0, 0, 0)$ we shall discuss the term $x_1(t)$ in (3.1). It must be (as in Case II) a periodic solution of

(3.11) $$L(y) = f(t, x_0(t), x_0'(t), 0)$$

where $L(y)$ is defined in (3.2). A necessary and sufficient condition that (3, 11) have a periodic solution is that $f(t, x_0(t), x_0'(t), 0)$ be orthogonal on $(0, L)$ to the periodic solution of the homogeneous equation adjoint to (3.2). But in Case III this periodic solution is $p(t)/W(t)$, so that the orthogonality condition condition is exactly (3.62). Assuming this condition to be fulfilled, the most general periodic solution of (3.11) is

(3.65) $$y = Cp(t) + \frac{B_1 q(t)}{KL} + \int_0^t \{\frac{p(s)q(t) - p(t)q(s)}{W(s)} f(s, x_0(s), x_0'(s), 0)\}ds.$$

In (3.65), C is an arbitrary constant. The integral is the partial derivative $x_\varepsilon(t, 0, 0, 0)$, and we shall use this notation later. When $C = 0$, we obtain the unique periodic solution of (3.11) which has $y'(0) = 0$. We denote this by $h_1(t)$, so that (3.65) can be written

(3.66) $$y = Cp(t) + h_1(t).$$

Because the first derivatives of F at $(0, 0, 0)$ are all zero, the only second partial derivatives that we shall need are those of F at $(0, 0, 0)$. These have the values:

$$F_{\xi\xi} = g(A, 0) \int_0^L \frac{p(s)}{W(s)} [- g_{xx}(x_0(s), x_0'(s))q(s)^2$$

$$- 2g_{xx'}(x_0(s), x_0'(s))q(s)q'(s) - g_{x'x'}(x_0(s), x_0'(s))q'(s)^2]ds,$$

$$F_{\xi\eta} = KLg(A, 0),$$

$$F_{\eta\eta} = 0$$

$$F_{\xi\varepsilon} = \int_0^L \frac{p(s)}{W(s)} [- g_{xx}(x_0(s), x_0'(s))q(s)x_\varepsilon(s)$$

(3.67)

$$- g_{xx'}(x_0(s), x_0'(s))[q(s)x_\varepsilon'(s) + q'(s)x_\varepsilon(s)]$$

$$- g_{x'x'}(x_0(s), x_0'(s))q'(s)x_\varepsilon'(s) + f_x(s, x_0(s), x_0'(s), 0)q(s)$$

$$+ f_{x'}(s, x_0(s), x_0'(s), 0)q'(s)]ds$$

$$F_{\eta\varepsilon} = -B_1 + \frac{1}{g(A,0)^2} \int_0^L \frac{p(s)}{W(s)} f_t(s, x_0(s), x_0'(s), 0)\,ds,$$

$$F_{\varepsilon\varepsilon} = \frac{1}{g(A,0)} \int_0^L \frac{p(s)}{W(s)} [-g_{xx}(x_0(s), x_0'(s))x_\varepsilon(s)^2$$

(3.67) (continued)
$$- 2g_{xx'}(x_0(s), x_0'(s))x_\varepsilon(s)x_\varepsilon'(s) - g_{x'x'}(x_0(s), x_0'(s))x_\varepsilon'(s)^2$$

$$+ 2f_x(s, x_0(s), x_0'(s), 0)x_\varepsilon(s) + 2f_{x'}(s, x_0(s), x_0'(s), 0)x_\varepsilon'(s)$$

$$+ 2f_\varepsilon(s, x_0(s), x_0'(s), 0)]\,ds.$$

Using the results in (3.67) we find the various partial derivatives of $J(\eta, \varepsilon)$ at $(0, 0)$ to be

(3.68)
$$J_\eta(0,0) = 0, \qquad J_\varepsilon(0,0) = 0, \qquad J_{\eta\eta}(0,0) = 0,$$

$$J_{\eta\varepsilon} = \frac{1}{g(A,0)^2} \int_0^L \frac{p(s)}{W(s)} f_t(s, x_0(s), x_0'(s), 0)\,ds,$$

$$J_{\varepsilon\varepsilon} = \frac{1}{g(A,0)} \int_0^L \frac{p(s)}{W(s)} \{-g_{xx}(x_0(s), x_0'(s))h_1(s)^2$$

$$- 2g_{xx'}(x_0(s), x_0'(s))h_1(s), h_1'(s) - g_{x'x'}(x_0(s), x_0'(s))h_1'(s)^2$$

$$+ 2f_x(s, x_0(s), x_0'(s), 0)h_1(s) + 2f_{x'}(s, x_0(s), x_0'(s), 0)h_1'(s)$$

$$+ 2f_\varepsilon(s, x_0(s), x_0'(s), 0)\}\,ds.$$

Hence near $\eta = \varepsilon = 0$, we have

$$J(\eta, \varepsilon) = J_{\eta\varepsilon}\eta\varepsilon + \frac{1}{2}J_{\varepsilon\varepsilon}\varepsilon^2 + \text{higher order terms}.$$

If the coefficient $J_{\eta\varepsilon}$ is not zero, we may solve $J(\eta, \varepsilon) = 0$ for η as a function of ε with

(3.69)
$$\eta = 0, \qquad \frac{d\eta}{d\varepsilon} = -\frac{J_{\varepsilon\varepsilon}}{2J_{\eta\varepsilon}}$$

at $\varepsilon = 0$, by the modification of the implicit function theorem mentioned earlier.

Thus when $J_{\eta\varepsilon} \neq 0$, we have shown that the equations $F = G = 0$ can be solved for ξ and η as functions of ε near $\varepsilon = 0$ with derivatives $d\xi/d\varepsilon$ and $d\eta/d\varepsilon$ at $\varepsilon = 0$ given by (3.64) and (3.69) respectively.

As in Case II we may also determine the values of $x_1(0)$ and $x_1'(0)$ by formal manipulation of the perturbation series. The orthogonality condition for the existence of the periodic term $x_2(t)$ gives an equation for C_1, where

$$x_1(t) = C_1 p(t) + h_1(t).$$

This equation for C_1 turns out to be

$$-g(A, 0)^2 J_{\eta\varepsilon} C_1 + \frac{1}{2} g(A, 0) J_{\varepsilon\varepsilon} = 0$$

which gives the value

$$C_1 = \frac{J_{\varepsilon\varepsilon}}{2g(A, 0) J_{\eta\varepsilon}}$$

which checks with the value of $x_1'(0)$ found from $d\eta/d\varepsilon$.

The foregoing results can be summarized in the following theorem.

THEOREM 3.7. *In equation* (1.2) *let* $g(x, x') \in C^2$ *and* $f(t, x, x', \varepsilon) \in C^1$. *Let equation* (1.1) *have the nonconstant periodic solution* $x_0(t)$ *of least period* L_0. *Let* $f(t, x, x', \varepsilon)$ *have least period* T *in* t, *where* T *is a rational multiple of* L_0. *Let* L *be the least common multiple of* T *and* L_0. *Let* $x_0(0) = A, x_0'(0) = 0$. *Further let*

$$\int_0^L g_{x'}(x_0(t), x_0'(t))\, dt = 0, \tag{3.70}$$

$$\int_0^L \frac{p(s)}{W(s)} f(s, x_0(s), x_0'(s), 0)\, ds = 0, \tag{3.71}$$

$$\int_0^L \frac{p(s)}{W(s)} f_t(s, x_0(s), x_0'(s), 0)\, ds \neq 0. \tag{3.72}$$

Finally, let the variation equation (3.2) *have no periodic solutions other than constant multiples of* $p(t)$.

Then (1.2) *has a unique periodic solution of period* L *of the form*

$$x = x_0(t) + \varepsilon x_1(t) + o(\varepsilon) \tag{3.73}$$

where $x_1(t)$ *is a uniquely determined solution of equation* (3.11) *namely*

$$x_1(t) = \frac{J_{\varepsilon\varepsilon}\, p(t)}{2J_{\eta\varepsilon}\, g(A, 0)} + h_1(t). \tag{3.74}$$

Remarks.

1. The condition (3.72) is exactly equivalent to the nonvanishing of the critical Jacobian in the theory of Coddington and Levinson or of Friedrichs.

2. When condition (3.71) does not hold, it may be that an appropriate translation of $x_0(t)$ will furnish a periodic solution. This will be investigated later in this section.

3. When condition (3.72) is not satisfied, it is necessary to use higher derivatives of F and G at (0, 0, 0) to find the period solution or solutions. We investigate one case of this in the present section.

The existence theory in the case where we use the third derivatives of F and G at (0, 0, 0) is very similar to that of Case II. Since the details are so similar we shall omit most details.

The function $x_2(t)$ of the perturbation series will surely exist if all the second partial derivatives of $J(\eta, \varepsilon)$ at (0, 0) vanish. Indeed, just as in Case II we may write

$$x_2(t) = \frac{1}{2} C_1^2 \phi_2(t) + C_1 k_2(t) + h_2(t) + C_2 p(t), \tag{3.75}$$

where ϕ_2, k_2, and h_2 are solutions of (3.25), (3.26), and (3.27) respectively with the same initial conditions. The only difference is that $h_1(t)$ is the $h_1(t)$ of the present section.

The formulas for the third derivatives of $J(\eta, \varepsilon)$ at (0, 0) are very similar to those of (3.29). Indeed we have

$$J_{\eta\eta\eta} = 0,$$

$$J_{\eta\eta\varepsilon} = \frac{1}{g(A, 0)^3} \int_0^L \frac{p(s)}{W(s)} f_{tt}(s, x_0(s), x_0'(s), 0)\, ds,$$

$$\begin{aligned} J_{\eta\varepsilon\varepsilon} = \frac{2}{g(A, 0)^2} \int_0^L \frac{p(s)}{W(s)} \{ & - g_{xx}(x_0(s), x_0'(s)) h_1(s)(h_1'(s) - k_2(s)) \\ & - g_{xx'}(x_0(s), x_0'(s)) [h_1(s)(h_1''(s) - k_2'(s)) + h_1'(s) \\ & (h_1'(s) - k_2)s))] - g_{x'x'}(x_0(s), x_0'(s)) h_1'(s) \end{aligned} \tag{3.76}$$

(3.76) (continued)

$$(h_1''(s) - k_2'(s)) + f_x(s, x_0(s), x_0'(s), 0)(h_1'(s) - k_2(s))$$

$$+ f_{x'}(s, x_0(s), x_0'(s), 0)(h_1''(s) - k_2'(s))$$

$$+ f_{tx}(s, x_0(s), x_0'(s), 0) h_1(s) + f_{tx'}(s, x_0(s), x_0'(s), 0) h_1'(s)$$

$$+ f_{t\varepsilon}(s, x_0(s), x_0'(s), 0)\} ds,$$

$$J_{\varepsilon\varepsilon\varepsilon} = \frac{6}{g(A, 0)} \int_0^L \frac{p(s)}{W(s)} \{ - g_{xx}(x_0(s), x_0'(s)) h_1(s) h_2(s)$$

$$- g_{xx'}(x_0(s), x_0'(s))[h_1(s) h_2'(s) + h_1'(s) h_2(s)]$$

$$- g_{x'x'}(x_0(s), x_0'(s)) h_1'(s) h_2'(s) - \frac{1}{6} g_{xxx}(x_0(s), x_0'(s)) h_1(s)^3$$

$$- \frac{1}{2} g_{xxx'}(x_0(s), x_0'(s)) h_1(s)^2 h_1'(s)$$

$$- \frac{1}{2} g_{xx'x'}(x_0(s), x_0'(s)) h_1(s) h_1'(s)^2$$

$$- \frac{1}{6} g_{x'x'x'}(x_0(s), x_0'(s)) h_1'(s)^3 + f_x(s, x_0(s), x_0'(s), 0) h_2(s)$$

$$+ f_{x'}(s, x_0(s), x_0'(s), 0) h_2'(s) + \frac{1}{2} f_{xx}(s, x_0(s) x_0'(s), 0) h_1(s)^2$$

$$+ f_{xx'}(s, x_0(s), x_0'(s), 0) h_1(s) h_1'(s)$$

$$+ f_{x'x'}(s, x_0(s), x_0'(s), 0) h_1'(s)^2 + f_{x\varepsilon}(s, x_0(s), x_0'(s), 0) h_1(s)$$

$$+ f_{x'\varepsilon}(s, x_0(s), x_0'(s), 0) h_1'(s) + \frac{1}{2} f_{\varepsilon\varepsilon}(s, x_0(s), x_0'(s), 0)\} ds.$$

The only difference between the formulas in (3.76) and (3.29) is the presence of $p(s)$ instead of $q(s)$ at the beginning of the integrands, and an extra factor of $g(A, 0)^{-2}$.

The discussion now proceeds as in Case II. In particular we find the partial derivative $d\eta/d\varepsilon$ at $\varepsilon = 0$ as a root of equation (3.31), assuming this equation has simple real roots. Of course its coefficients are this time taken from (3.76). Also we can find the coefficient C_1 in $x_1(t)$ by using the perturbation series. The equation for C_1 again turns out to be (3.38).

THEOREM 3.8. *In equation* (1.2) *let* $g(x, x') \in C^3$ *and* $f(t, x, x', \varepsilon) \in C^2$. *Let all the hypotheses of Theorem* 3.7 *be satisfied except for* (3.72). *Let*

(3.77) $$\int_0^L \frac{p(s)}{W(s)} f(s, x_0(s), x_0'(s), 0)\, ds = 0$$

(3.78) $$\int_0^L \frac{p(s)}{W(s)} f_t(s, x_0(s), x_0'(s), 0)\, ds = 0,$$

(3.79) $$J_{\varepsilon\varepsilon}(0, 0) = 0$$

(3.80) $$J_{\eta\varepsilon\varepsilon}(0, 0)^2 - \frac{4}{3} J_{\eta\eta\varepsilon}(0, 0) J_{\varepsilon\varepsilon\varepsilon}(0, 0) > 0.$$

Then (1.2) *has one or two periodic solutions of period* L *of the form*

(3.81) $$x = x_0(t) + \varepsilon x_1(t) + o(\varepsilon)$$

where $x_1(t)$ *is a uniquely determined solution of* (3.11) *according as*

(3.82) $$\int_0^L \frac{p(s)}{W(s)} f_{tt}(s, x_0(s), x_0'(s), 0)\, ds$$

does or does not vanish.

For the stability investigation we proceed as in Case II so far as the successive approximation solution of the variation equation is concerned. In Case III it is found that

(3.83)
$$\begin{aligned} \phi(L) &= 1 + \varepsilon \int_0^L \frac{p(s)}{W(s)} \{-g_{xx}(x_0(s), x_0'(s)) x_1(s) q(s) \\ &\quad - g_{xx'}(x_0(s), x_0'(s)) [x_1'(s) q(s) + x_1(s) q'(s)] \\ &\quad - g_{x'x'}(x_0(s), x_0'(s)) x_1'(s) q'(s) + f_x(s, x_0(s), x_0'(s), 0) q(s) \\ &\quad + f_{x'}(s, x_0(s), x_0'(s), 0) q'(s)\} ds + o(\varepsilon), \\ &= 1 + \varepsilon d_{11} + o(\varepsilon), \end{aligned}$$

$$\phi'(L) = g(A, 0) q'(L) + \varepsilon d_{21} + o(\varepsilon) = -KLg(A, 0)^2 + \varepsilon d_{21} + o(\varepsilon),$$

(3.83) (continued)

$$\psi(L) = -\varepsilon \frac{1}{g(A,0)^2} \int_0^L \frac{p(s)}{W(s)} \{-g_{xx}(x_0(s), x_0'(s))x_1(s)p(s)$$

$$- g_{xx'}(x_0(s), x_0'(s))[x_1'(s)p(s) + x_1(s)p'(s)]$$

$$- g_{x'x'}(x_0(s), x_0'(s))x_1'(s)p'(s) + f_x(s, x_0(s), x_0'(s), 0)p(s)$$

$$+ f_{x'}(s, x_0(s), x_0'(s), 0)p'(s)\} ds + o(\varepsilon)$$

$$= \varepsilon d_{12} + o(\varepsilon),$$

$$\psi'(L) = 1 + \varepsilon \int_0^L [KL \frac{p(s)}{W(s)} - \frac{q(s)}{W(s)}] \{-g_{xx}(x_0(s), x_0'(s))x_1(s)p(s)$$

$$- g_{xx'}(x_0(s), x_0'(s))[x_1'(s)p(s) + x_1(s)p'(s)]$$

$$- g_{x'x'}(x_0(s), x_0'(s))x_1'(s)p'(s) + f_x(s, x_0(s)x_0'(s), 0)p(s)$$

$$+ f_{x'}(s, x_0(s), x_0'(s), 0)p'(s)\} ds + o(\varepsilon)$$

$$= 1 + \varepsilon d_{22} + o(\varepsilon).$$

The value of d_{21} is of no significance in Theorem 3.9.

The characteristic multipliers are the eigenvalues of

(3.84)
$$\begin{Vmatrix} \phi(L) & \psi(L) \\ \phi'(L) & \psi'(L) \end{Vmatrix}.$$

The characteristic equation of the matrix (3.84) is

(3.85)
$$(\lambda - 1)^2 - (d_{11} + d_{22})(\lambda - 1)\varepsilon + \varepsilon KLg(A,0)^2 d_{12} + o(\varepsilon) = 0$$

Hence if $\varepsilon KLg(A,0)^2 d_{12}$ is negative, the two roots of (3.85) will be real and separated by 1 for small ε. On the other hand, if $\varepsilon KLg(A,0)^2 d_{12}$ is positive, the roots will be imaginary for small ε. In this latter case, the square of their common absolute value will be the determinant of the matrix (3.84) which reduces to

(3.86)
$$1 + \varepsilon(d_{11} + d_{22} + KLg(A,0)^2 d_{12}) + o(\varepsilon)$$

Both these critical expressions can be simplified. In fact

(3.87) $$d_{12} = \frac{1}{g(A,0)^2} \int_0^L \frac{p(s)}{W(s)} f_t(s, x_0(s), x_0'(s), 0)\, ds,$$

while

$$d_{11} + d_{22} + KLg(A,0)^2 d_{12}$$

$$= \int_0^L \frac{p(s)q'(s) - q(s)p'(s)}{W(s)} \{-g_{xx'}(x_0(s), x_0'(s))x_1(s)$$

$$- g_{x'x'}(x_0(s), x_0'(s))x_1'(s) + f_{x'}(s, x_0(s), x_0'(s), 0)\}\, ds.$$

On account of the definition of $W(s)$, the first factor in the integrand above is 1. Also certain other terms will cancel and we obtain finally

(3.88) $$d_{11} + d_{22} + KLg(A,0)^2 d_{12} = \int_0^L \{-g_{xx'}(x_0(s), x_0'(s))h_1(s)$$

$$- g_{x'x'}(x_0(s), x_0'(s))h_1'(s) + f_{x'}(s, x_0(s), x_0'(s), 0)\}\, ds.$$

THEOREM 3.9. *Let all the hypotheses of Theorem* 3.7 *be satisfied. Then the periodic solution found in that theorem has the following stability properties:*

(a) *If*

(3.89) $$\varepsilon K \int_0^L \frac{p(s)}{W(s)} f_t(s, x_0(s), x_0'(s), 0)\, ds$$

is negative, the solution is directly unstable.

(b) *If* (3.89) *is positive, the solution is completely stable or completely unstable according as*

(3.90) $$\varepsilon \int_0^L \{-g_{xx'}(x_0(s), x_0'(s))h_1(s) - g_{x'x'}(x_0(s), x_0'(s))h_1'(s)$$

$$+ f_{x'}(s, x_0(s), x_0'(s), 0)\}\, ds.$$

is negative or positive.

Proof of Theorem 3.9. The statement in (a) follows from the remark following equation (3.85). The statement in (b) follows from this remark, and equations (3.86) and (3.87). This completes the proof.

Remark. In case (3.89) is positive and (3.90) is zero, information can at times be gained from the third derivatives of F and G. We investigate this later in the section.

When (3.89) is zero, the hypothesis of Theorem 3.7 is not fulfilled. In cases where the hypothesis of Theorem 3.8 is filled, we are able to investigate stability of the periodic solutions determined by it.

As in Section 3.3, we carry the successive approximation solution of the variation equation (3.39) a second step. We obtain

$$(3.91)\qquad \begin{aligned} \phi(L) &= 1 + \varepsilon d_{11} + \varepsilon^2 f_{11} + o(\varepsilon^2), \\ \phi'(L) &= -KLg(A,0)^2 + \varepsilon d_{21} + \varepsilon^2 f_{21} + o(\varepsilon^2), \\ \psi(L) &= \varepsilon d_{12} + \varepsilon^2 f_{12} + o(\varepsilon^2), \\ \psi'(L) &= 1 + \varepsilon d_{22} + \varepsilon^2 d_{22} + o(\varepsilon^2). \end{aligned}$$

Under the hypotheses of Theorem 3.8, $d_{12} = 0$. The characteristic equation of the matrix (3.84) is

$$(3.92)\qquad \begin{aligned} &(\lambda-1)^2 + [\varepsilon(d_{11}+d_{22}) + \varepsilon^2(f_{11}+f_{22})](\lambda-1) \\ &\qquad + \varepsilon^2(d_{11}d_{22} + KLg(A,0)^2 f_{12}) + o(\varepsilon^2) = 0 \end{aligned}$$

Hence form small nonzero ε, the roots of (3.92) will be real and separated by 1 if $d_{11}d_{22} + KLg(A,0)^2 f_{12}$ is negative, and will be both real and not separated by 1 or else both imaginary if this quantity is positive. In this latter case the product of the two roots is the determinant of the matrix (3.84), which is

$$1 + \varepsilon(d_{11} + d_{22}) + o(\varepsilon)$$

under the hypotheses of Theorem 3.8. This again gives the condition (3.90) as in Theorem 3.9.

The quantity $d_{11}d_{22} + KLg(A,0)^2 f_{12}$ can be simplified. The computation is long and tedious, and the final result is

$$(3.93)\qquad d_{11}d_{22} + KLg(A,0)^2 f_{12} = -KLC_1 g(A,0)^3 J_{\eta\eta\varepsilon} + \frac{1}{2} KLg(A,0)^2 J_{\eta\varepsilon\varepsilon}.$$

Note that apart from a factor of $-KLg(A,0)^2$, this is the derivative with respect to C_1 of the left member of equation (3.38), and hence is not zero, since C_1 is assumed to be a simple real root of (3.38).

THEOREM 3.10. *Let all of the hypotheses of Theorem* 3.8 *be satisfied. Then the periodic solutions found in that theorem have the following stability properties.*

(a) *If*

$$(3.94) \qquad - KC_1 g(A, 0) J_{\eta\eta\varepsilon} + \frac{1}{2} KJ_{\eta\varepsilon\varepsilon}$$

is negative, the solution is directly unstable.

(b) *If* (3.94) *is positive, the solution is completely stable or completely unstable according as*

$$(3.90) \qquad \varepsilon \int_0^L \{- g_{xx'}(x_0(s), x_0'(s)) h_1(s) - g_{x'x'}(x_0(s), x_0'(s)) h_1'(s) + f_{x'}(s, x_0(s), x_0'(s), 0)\} ds$$

is negative or positive.

Proof of Theorem 3.10. The proof is very similar to that of Theorem 3.9. The quantity (3.94) is obtained from (3.93) by suppressing the positive factor $Lg(A, 0)^2$. The statement in (b) is the same as in Theorem 3.9. This completes the proof.

In both the preceding theorems, the quantity (3.90) appeared and was used to distinguish between complete stability and complete instability. In case this quantity is zero, it is necessary to pursue the investigation further. Fortunately if we know the third derivatives of $g(x, x')$ and the second derivatives of $f(t, x, x', \varepsilon)$ and we know the solution $x(t)$ as far as $x_0(t) + \varepsilon x_1(t) + \varepsilon^2 x_2(t) + o(\varepsilon^2)$, it is possible to obtain a substitute expression for (3.90) in case (3.90) is zero. The determinant of the matrix (3.84), when (3.90) is zero is:

$$(3.95) \qquad 1 + \varepsilon^2 (f_{11} + f_{22} + KLg(A, 0)^2 f_{12} + d_{11} d_{22} - d_{12} d_{21}) + o(\varepsilon^2),$$

Thus an effective substitute for (3.90) would be

$$f_{11} + f_{22} + KLg(A, 0)^2 f_{12} + d_{11} d_{22} - d_{12} d_{21} .$$

After a rather lengthy computation, this is found to reduce to

$$(3.96) \qquad \int_0^L \{ - g_{xx'}(x_0(s), x_0'(s)) x_2(s) - g_{x'x'}(x_0(s), x_0'(s)) x_2'(s) - \frac{1}{2} g_{xxx'}(x_0(s), x_0'(s)) x_1(s)^2 - g_{xx'x'}(x_0(s), x_0'(s)) x_1(s) x_1'(s)$$

$$- \frac{1}{2} g_{x'x'x'}(x_0(s), x_0'(s)) x_1'(s)^2 + f_{xx'}(s, x_0(s), x_0'(s), 0) x_1(s)$$

$$+ f_{x'x'}(s, x_0(s), x_0'(s), 0) x_1'(s) + f_{x'\varepsilon}(s, x_0(s), x_0'(s), 0) \} ds.$$

In (3.96), we understand that $x_1(s) = h_1(s) + C_1 p(s)$. Assuming that $J_{\eta\varepsilon} \neq 0$, we can find $x_2(s)$ directly from the perturbation series by solving a nonhomogeneous linear equation. In case that $J_{\eta\varepsilon}$ and $J_{\varepsilon\varepsilon}$ are both zero, we understand that $x_2(s) = \frac{1}{2} C_1^2 \phi_2(s) + C_1 k_2(s) + h_2(s) + C_2 p(s)$. Certain terms in (3.96) are found to give zero, in particular those involving C_1^2 and C_2. We now have the following theorem.

THEOREM 3.11. *Let* $g(x, x') \in C^3$, $f(t, x, x', \varepsilon) \in C^2$. *If in Theorem 3.9 or Theorem 3.10 the quantity* (3.90) *is zero, it may be replaced by* (3.96) *provided that* (3.96) *is not zero.*

We now turn to the case in which the periodic solution sought is a perturbation of a translation $x_0(t+\tau)$ of $x_0(t)$. The theory is very similar to that in Section 3.2, so we shall make the discussion comparatively brief.

Let $F_1(\tau)$ be defined by

$$F_1(\tau) = \int_0^L \frac{p(t)}{W(t)} f(t-\tau, x_0(t), x_0'(t), 0)\, dt. \tag{3.97}$$

We note that

$$F_1'(\tau) = -\int_0^L \frac{p(t)}{W(t)} f_t(t-\tau, x_0(t), x_0'(t), 0)\, dt, \tag{3.98}$$

and that

$$F_1''(\tau) = \int_0^L \frac{p(t)}{W(t)} f_{tt}(t-\tau, x_0(t), x_0'(t), 0)\, dt. \tag{3.99}$$

As in Section 3.2, a necessary condition for the existence of the periodic solution is that $F_1(\tau) = 0$. Assuming that $F_1(\tau) = 0$, let $h_1(t, \tau)$ be that periodic solution of

$$L(y) = f(t-\tau, x_0(t), x_0'(t), 0) \tag{3.100}$$

for which $y'(0) = 0$. Now let $G_1(\tau)$ be defined by

$$(3.101) \qquad G_1(\tau) = \int_0^L [-g_{xx'}(x_0(s), x_0'(s))h_1(s,\tau)$$
$$- g_{x'x'}(x_0(s), x_0'(s))h_1'(s,\tau) + f_{x'}(s-\tau, x_0(s), x_0'(s), 0)]\, ds.$$

The prime on $h_1(t,\tau)$ denotes differentiation with respect to t.

Similarly let $G_2(\tau)$ be defined from (3.96) where it is understood that $x_1(t)$ and $x_2(t)$ are computed using $f(t-\tau, x, x', \varepsilon)$, and that the first argument in f and its derivatives is $s-\tau$.

Using these functions, we can state the following theorem equivalent to Theorems 3.7, 3.9, and 3.11.

THEOREM 3.12. *Let all of the hypotheses of Theorem* 3.7 *preceding equation* (3.71) *hold, together with the hypothesis that the variation equation has but one periodic solution. Let* $F_1(\tau) = 0$, *and let* $F_1'(\tau) \neq 0$. *Then there exists for small nonzero* ε *a periodic solution of*

$$(3.102) \qquad x'' + g(x, x') = \varepsilon f(t-\tau, x, x', \varepsilon)$$

of the form

$$(3.103) \qquad x = x_0(t) + \varepsilon x_1(t) + o(\varepsilon)$$

and correspondingly a periodic solution of

$$(3.104) \qquad x'' + g(x, x') = \varepsilon f(t, x, x', \varepsilon)$$

of the form

$$(3.105) \qquad x = x_0(t+\tau) + \varepsilon x_1(t+\tau) + o(\varepsilon).$$

If $\varepsilon KF_1'(\tau)$ *is positive, the solution is directly unstable. If* $\varepsilon KF_1'(\tau)$ *is negative, the solution is completely stable or unstable according as* $\varepsilon G_1(\tau)$ *is negative or positive. In case* $G_1(\tau) = 0$, *and* $g(x, x') \in C^3$, $f(t, x, x', \varepsilon) \in C^2$ *the solution is completely stable or completely unstable according as* $G_2(\tau)$ *is negative or positive.*

Remarks hold similar to those following Theorem 3.5.

In case $F_1(\tau) \equiv 0$, we can obtain a theorem equivalent to theorems 3.8, 3.10, and 3.11. For this we define $F_2(\tau)$ as being $J_{\varepsilon\varepsilon}(0,0)$, where $h_1(s)$ is replaced by $h_1(s, \tau)$, and the first argument of all derivatives of f is $s - \tau$. We also omit the factor $1/g(A, 0)$. As in Section 3.2 it is found that $F_2'(\tau) = -J_{\eta\varepsilon\varepsilon}$ apart from a positive factor, $1/(g(A, 0)^2)$.

THEOREM 3.13. *Let all of the hypotheses about continuity of Theorem* 3.8 *hold, and let all the hypotheses of Theorem* 3.7 *preceding equation* (3.71) *as well as the hypothesis that the variation equation has but one periodic solution hold. Let* $F_1(\tau) \equiv 0$, *let* $F_2(\tau) = 0$, *and let* $F_2'(\tau) \neq 0$.

Then there exists a periodic solution of

(3.102) $$x'' + g(x, x') = \varepsilon f(t - \tau, x, x', \varepsilon)$$

of the form

(3.103) $$x = x_0(t) + \varepsilon x_1(t) + o(\varepsilon),$$

and correspondingly a periodic solution of

(3.104) $$x'' + g(x, x') = \varepsilon f(t, x, x', \varepsilon)$$

of the form

(3.105) $$x = x_0(t + \tau) + \varepsilon x_1(t + \tau) + o(\varepsilon).$$

If $KF_2'(\tau)$ *is positive, the solution is directly unstable. If* $KF_2'(\tau)$ *is negative, the solution is completely stable or completely unstable according as* $\varepsilon G_1(\tau)$ *is negative or positive, and in case* $G_1(\tau) = 0$, *according as* $G_2(\tau)$ *is negative or positive.*

Appendix to Section 3.3

In this appendix we derive the expressions for some of the partial derivatives of F and G at $(0, 0, 0)$. In the following, all derivatives of $x(t, \xi, \eta, \varepsilon)$ are understood to be evaluated at $\xi = \eta = \varepsilon = 0$. All derivatives of $g(x, x')$ are evaluated at $x = x_0(t)$, $x' = x_0'(t)$. All derivatives of $f(t, x, x', \varepsilon)$ are evaluated at $x = x_0(t)$, $x' = x_0'(t)$, $\varepsilon = 0$.

The partial derivatives x_ξ, x_η, and x_ε satisfy

$$x_\xi'' + g_{x'}x_\xi' + g_x x_\xi = 0 \qquad x_\xi(0) = 1, \qquad x_\xi'(0) = 0,$$

$$x_\eta'' + g_{x'}x_\eta' + g_x x_\eta = 0 \qquad x_\eta(0) = 0, \qquad x_\eta'(0) = 1,$$

$$x_\varepsilon'' + g_{x'}x_\varepsilon' + g_x x_\varepsilon = f \qquad x_\varepsilon(0) = 0, \qquad x_\varepsilon'(0) = 0.$$

As a result,

$$x_\xi = g(A,0)\,q(t), \qquad x_\eta = -\frac{p(t)}{g(A,0)}$$

$$x_\varepsilon = \int_0^t \frac{p(s)q(t) - q(s)p(t)}{W(s)} f(s, x_0(s), x_0'(s), 0)\,ds.$$

Using the known values (3.4) and (3.59), we find

$$F_\xi = x_\xi(L) - 1 = g(A,0)\,q(L) - 1 = 0$$

$$G_\xi = x_\xi'(L) = g(A,0)\,q'(L) = -KLg(A,0)^2$$

$$F_\eta = x_\eta(L) = -\frac{p(L)}{g(A,0)} = 0$$

$$G_\eta = x_\eta'(L) - 1 = -\frac{p'(L)}{g(A,0)} - 1 = 0$$

$$F_\varepsilon = x_\varepsilon(L) = \int_0^L \frac{p(s)q(L) - q(s)p(L)}{W(s)} f(s, x_0(s), x_0'(s), 0)\,ds$$
$$= \frac{1}{g(A,0)} \int_0^L \frac{p(s)}{W(s)} f(s, x_0(s), x_0'(s), 0)\,ds,$$

$$G_\varepsilon = x_\varepsilon'(L) = \int_0^L \frac{p(s)q'(L) - q(s)p'(L)}{W(s)} f(s, x_0(s), x_0'(s), 0)\,ds$$
$$= -KLg(A,0) \int_0^L \frac{p(s)}{W(s)} f(s, x_0(s), x_0'(s), 0)\,ds$$
$$+ g(A,0) \int_0^L \frac{q(s)}{W(s)} f(s, x_0(s), x_0'(s), 0)\,ds.$$

The above are the results given in (3.61).

The second derivatives of x satisfy linear second order differential equations. We shall evaluate $F_{\xi\eta}$ and $F_{\eta\varepsilon}$, which are the values at $t = L$ of $x_{\xi\eta}$ and $x_{\eta\varepsilon}$ respectively. We have

$$x''_{\xi\eta} + g_{x'}x'_{\xi\eta} + g_x x_{\xi\eta} = -g_{xx}x_\xi x_\eta - g_{xx'}x_\xi x'_\eta - g_{xx'}x'_\xi x_\eta - g_{x'x'}x'_\xi x'_\eta$$

$$x_{\xi\eta}(0) = 0, \qquad x'_{\xi\eta}(0) = 0.$$

Hence

$$x_{\xi\eta} = \int_0^t \frac{p(s)q(t) - q(s)p(t)}{W(s)} \{ g_{xx}(x_0(s), x_0'(s))p(s)q(s)$$

$$+ g_{xx'}(x_0(s), x_0'(s))[p'(s)q(s) + p(s)q'(s)]$$

$$+ g_{x'x'}(x_0(s), x_0'(s))p'(s)q'(s)\} ds$$

and

$$F_{\xi\eta} = x_{\xi\eta}(L) = \frac{1}{g(A, 0)} \int_0^L \frac{p(s)}{W(s)} [g_{xx}(x_0(s), x_0'(s)p(s)q(s)$$

$$+ g_{xx'}(x_0(s), x_0'(s))\{p'(s)q(s) + p(s)q'(s)\}$$

$$+ g_{x'x'}(x_0(s), x_0'(s))p'(s)q'(s)] ds$$

The integrand can be verified to be the derivative of

$$\frac{1}{W(s)}[p'(s)q'(s) + g_x(x_0(s), x_0'(s))p(s)q(s)$$

$$+ g_{x'}(x_0(s), x_0'(s))p(s)q'(s)],$$

using the fact that $p(s) \equiv x_0'(s)$, and the differential equations satisfied by $p(s)$, $q(s)$, and $W(s)$. Then since $p(0) = p(L) = q'(0) = 0$, and $W(L) = 1$, we finally have

$$F_{\xi\eta} = \frac{1}{g(A, 0)} p'(L)q'(L) = KLg(A, 0),$$

which is the value given in (3.67).

The partial derivative $x_{\eta\varepsilon}$ satisfies

$$x''_{\eta\varepsilon} + g_{x'}x'_{\eta\varepsilon} + g_x x_{\eta\varepsilon} = -g_{xx}x_\eta x_\varepsilon - g_{xx'}x_\eta x'_\varepsilon - g_{xx'}x'_\eta x_\varepsilon$$

$$- g_{x'x'}x'_\eta x'_\varepsilon + f_x x_\eta + f_{x'}x'_\eta.$$

By reasoning as above, we find

$$F_{\eta\varepsilon} = x_{\eta\varepsilon}(L) = \frac{1}{g(A,0)^2}\int_0^L \frac{p(s)}{W(s)}\{g_{xx}(x_0(s), x_0'(s)\, p(s)\, x_\varepsilon(s)$$

$$+ g_{xx'}(x_0(s), x_0'(s))[p(s)x_\varepsilon'(s) + p'(s)x_\varepsilon(s)]$$

$$+ g_{x'x'}(x_0(s), x_0'(s))\, p'(s)\, x_\varepsilon'(s) - f_x(s, x_0(s), x_0'(s), 0)\, p(s)$$

$$- f_{x'}(s, x_0(s), x_0'(s), 0)\, p'(s)\}\, ds$$

The integrand can be written in the form

$$\frac{p(s)}{W(s)} f_t(s, x_0(s), x_0'(s), 0) + \frac{d}{ds}\Big[\frac{1}{W(s)}(p'(s)x_\varepsilon'(s)$$

$$+ p(s)x_\varepsilon(s)\, g_x(x_0(s), x_0'(s)) + p(s)x_\varepsilon'(s)\, g_{x'}(x_0(s), x_0'(s))$$

$$- p(s) f(s, x_0(s), x_0'(s), 0))\Big].$$

Then since $p(0) = p(L) = x_\varepsilon'(0) = 0$, and $W(L) = 1$, we have

$$F_{\eta\varepsilon} = \frac{1}{g(A,0)^2}\left[\int_0^L \frac{p(s)}{W(s)} f_t(s, x_0(s), x_0'(s), 0)\, ds + p'(L)\, x_\varepsilon'(L)\right].$$

Since we are assuming $F_\varepsilon = 0$, $x_\varepsilon'(L) = G_\varepsilon = g(A,0)B_1$. Thus finally

$$F_{\eta\varepsilon} = \frac{1}{g(A,0)^2}\int_0^L \frac{p(s)}{W(s)} f_t(s, x_0(s), x_0'(s), 0)\, ds - B_1,$$

which is the value given in (3.67).

3.4. Existence and Stability in Case IV. According to Section 3.1, Case IV occurs if and only if all solutions of the variation equation (3.2) are periodic. Both characteristic multipliers are equal to +1.

We use as the fundamental set of solutions of (3.2), the periodic functions $p(t) \equiv x_0'(t)$, and $q(t)$, defined by the initial conditions $q(0) = 1/g(A,0)$, $q'(0) = 0$. We then have

$$p(0) = p(L) = 0 \qquad p'(0) = p'(L) = -g(A, 0)$$

(3.106)

$$q(0) = q(L) = \frac{1}{g(A, 0)} \qquad q'(0) = q'(L) = 0.$$

Because of the choice of initial values, the Wronskian

$$W(t) = p(t)q'(t) - p'(t)q(t)$$

satisfies $W(0) = 1$. $W(t)$ is periodic, and is given by

$$W(t) = e^{-\int_0^t g_{x'}(x_0(s), x_0'(s))\,ds}$$

The fundamental solutions $\phi(t)$ and $\psi(t)$ of (3.2) are given by

$$\phi(t) = g(A, 0)\,q(t), \qquad \psi(t) = -\frac{p(t)}{g(A, 0)}.$$

The first partial derivatives of F and G at (0, 0, 0) have the values

$$F_\xi = F_\eta = G_\xi = G_\eta = 0$$

$$F_\varepsilon = \frac{1}{g(A, 0)} \int_0^L \frac{p(s)}{W(s)} f(s, x_0(s), x_0'(s), 0)\,ds$$

$$G_\varepsilon = g(A, 0) \int_0^L \frac{q(s)}{W(s)} f(s, x_0(s), x_0'(s), 0)\,ds$$

The Jacobian of F and G with respect to ξ and η is zero, and the rank of the Jacobian matrix is zero. As a result, in order that $d\xi/d\varepsilon$ and $d\eta/d\varepsilon$ should exist at $\varepsilon = 0$, the augmented matrix

$$\begin{Vmatrix} F_\xi & F_\eta & F_\varepsilon \\ G_\xi & G_\eta & G_\varepsilon \end{Vmatrix}$$

must have rank zero, and this is equivalent to the two requirements

$$\int_0^L \frac{p(s)}{W(s)} f(s, x_0(s), x_0'(s), 0)\,ds = 0$$

(3.107)

$$\int_0^L \frac{q(s)}{W(s)} f(s, x_0(s), x_0'(s), 0)\,ds = 0.$$

Assuming (3.107) to hold, all six of the first partial derivatives of F and G are zero. To study the solution of the equations $F = G = 0$, we must proceed to the second partial derivatives of F and G.

There is one important case in which many of the partial derivatives of F and G vanish automatically. If the equation $x'' + g(x, x') = 0$ is such that every solution in the neighborhood of $x_0(t)$ is periodic with period L_0, Case IV is sure to occur. Thus for $\varepsilon = 0$, the function $x(t, \xi, \eta, \varepsilon)$ is periodic with period L for all choices of ξ and η near $\xi = \eta = 0$. As a result, $F(\xi, \eta, 0)$ and $G(\xi, \eta, 0)$ vanish identically in some neighborhood of $\xi = \eta = 0$. Hence all the partial derivatives of F and G at $(0, 0, 0)$ which do not involve differentiation with respect to ε will be zero in this special case. Note that the very important quasilinear case ($g(x, x') \equiv kx$ for some $k > 0$) is included in this.

The term $x_1(t)$ in the solution $x = x_0(t) + \varepsilon x_1(t) + o(\varepsilon)$ is itself a periodic solution of the differential equation

$$(3.108) \qquad \begin{aligned} &y'' + g_{x'}(x_0(t), x_0'(t))y' + g_x(x_0(t), x_0'(t))y \\ &\qquad = f(t, x_0(t), x_0'(t), 0). \end{aligned}$$

But (3.108) is the same equation as that satisfied by x_ε. Also, if conditions (3.107) are satisfied, every solution of (3.108) is periodic. Hence we have

$$x_1(t) = C_1 p(t) + C_1^* q(t) + x_\varepsilon(t),$$

where the constants C_1 and C_1^* are as yet undetermined. To have a notation similar to that in earlier sections, we denote $x_\varepsilon(t)$ also by $h_1(t)$.

The second partial derivatives of F and G at $(0, 0, 0)$ have the following values (if conditions (3.107) are satisfied).

$$F_{\xi\xi} = -g(A, 0) \int_0^L [g_{xx'}(x_0(s), x_0'(s))q(s) + g_{x'x'}(x_0(s), x_0'(s))q'(s)]ds,$$

$$(3.109) \qquad \begin{aligned} G_{\xi\xi} = -g(A, 0)^3 \int_0^L \frac{q(s)}{W(s)} &[g_{xx}(x_0(s), x_0'(s))q(s)^2 \\ &+ 2g_{xx'}(x_0(s), x_0'(s))q(s)q'(s) + g_{x'x'}(x_0(s), x_0'(s))q'(s)^2]ds \end{aligned}$$

$$F_{\xi\eta} = 0,$$

$$G_{\xi\eta} = 0,$$

$$F_{\eta\eta} = 0,$$

$$G_{\eta\eta} = 0,$$

$$\begin{aligned} F_{\xi\varepsilon} = \int_0^L [& -g_{xx'}(x_0(s), x_0'(s))h_1(s) - g_{x'x'}(x_0(s), x_0'(s))h_1'(s) \\ & + f_{x'}(s, x_0(s), x_0'(s), 0)]ds \\ & - \int_0^L \frac{q(s)}{W(s)} f_t(s, x_0(s), x_0'(s), 0)\,ds, \end{aligned}$$

(3.109) (continued)

$$\begin{aligned} G_{\xi\varepsilon} = g(A, 0) \int_0^L \frac{q(s)}{W(s)} [& - g_{xx}(x_0(s), x_0'(s))q(s)h_1(s) \\ & - g_{xx'}(x_0(s), x_0'(s))\{q(s)h_1'(s) + q'(s)h_1(s)\} \\ & - g_{x'x'}(x_0(s), x_0'(s))q'(s)h_1'(s) + f_x(s, x_0(s), x_0'(s), 0)q(s) \\ & + f_{x'}(s, x_0(s), x_0'(s), 0)q'(s)]ds \end{aligned}$$

$$F_{\eta\varepsilon} = \frac{1}{g(A, 0)^2} \int_0^L \frac{p(s)}{W(s)} f_t(s, x_0(s), x_0'(s), 0)\,ds,$$

$$G_{\eta\varepsilon} = \int_0^L \frac{q(s)}{W(s)} f_t(s, x_0(s), x_0'(s), 0)\,ds,$$

$$\begin{aligned} F_{\varepsilon\varepsilon} = \frac{1}{g(A, 0)} \int_0^L & \frac{p(s)h_1'(s) - p'(s)h_1(s)}{W(s)} [- g_{xx'}(x\ (s), x\ '(s))h_1(s) \\ & - g_{x'x'}(x_0(s), x_0'(s))h_1'(s) + 2f_{x'}(s, x_0(s), x_0'(s), 0)]ds \\ & - \frac{2}{g(A, 0)} \int_0^L \frac{h_1(s)}{W(s)} f_t(s, x_0(s), x_0'(s), 0)\,ds \\ & - \frac{1}{g(A, 0)} \int_0^L \frac{h_1(s)}{W(s)} g_{x'}(x_0(s), x_0'(s))f(s, x_0(s), x_0'(s), 0)\,ds \\ & + \frac{2}{g(A, 0)} \int_0^L \frac{q(s)}{W(s)} f_\varepsilon(s, x_0(s), x_0'(s), 0)\,ds, \end{aligned}$$

$$G_{\varepsilon\varepsilon} = g(A, 0) \int_0^L \frac{q(s)}{W(s)} [- g_{xx}(x_0(s), x_0'(s))h_1(s)^2$$

(3.109) (continued)
$$- 2g_{xx'}(x_0(s), x_0'(s))h_1(s)h_1'(s) - g_{x'x'}(x_0(s), x_0'(s))h_1'(s)^2$$
$$+ 2f_x(s, x_0(s), x_0'(s), 0)h_1(s) + 2f_{x'}(s, x_0(s), x_0'(s), 0)h_1'(s)$$
$$+ 2f_\varepsilon(s, x_0(s), x_0'(s), 0)]ds.$$

Since the first derivatives of F and G at $(0, 0, 0)$ all vanish, we have that near $(0, 0, 0)$

$$F(\xi, \eta, \varepsilon) = \frac{1}{2}F_{\xi\xi}\xi^2 + F_{\xi\varepsilon}\xi\varepsilon + F_{\eta\varepsilon}\eta\varepsilon + \frac{1}{2}F_{\varepsilon\varepsilon}\varepsilon^2$$
$$+ \text{higher order terms,}$$

(3.110)

$$G(\xi, \eta, \varepsilon) = \frac{1}{2}G_{\xi\xi}\xi^2 + G_{\xi\varepsilon}\xi\varepsilon + G_{\eta\varepsilon}\eta\varepsilon + \frac{1}{2}G_{\varepsilon\varepsilon}\varepsilon^2$$
$$+ \text{higher order terms.}$$

We seek to solve $F = G = 0$ for ξ and η as functions of ε with the derivatives $d\xi/d\varepsilon$ and $d\eta/d\varepsilon$ finite at $\varepsilon = 0$. In (3.110) set $\xi = u\varepsilon$, and $\eta = v\varepsilon$. We obtain

$$\varepsilon^2(\frac{1}{2}F_{\xi\xi}u^2 + F_{\xi\varepsilon}u + F_{\eta\varepsilon}v + \frac{1}{2}F_{\varepsilon\varepsilon}) + o(\varepsilon^2) = 0$$

(3.111)

$$\varepsilon^2(\frac{1}{2}G_{\xi\xi}u^2 + G_{\xi\varepsilon}u + F_{\eta\varepsilon}v + \frac{1}{2}G_{\varepsilon\varepsilon}) + o(\varepsilon^2) = 0.$$

Since a solution involving $\varepsilon \equiv 0$ does not interest us, we divide each equation in (3.111) by ε^2, and obtain two equations for u and v.

$$\frac{1}{2}F_{\xi\xi}u^2 + F_{\xi\varepsilon}u + F_{\eta\varepsilon}v + \frac{1}{2}F_{\varepsilon\varepsilon} + o(1) = 0$$

(3.112)

$$\frac{1}{2}G_{\xi\xi}u^2 + G_{\xi\varepsilon}u + G_{\eta\varepsilon}v + \frac{1}{2}G_{\varepsilon\varepsilon} + o(1) = 0.$$

Now let (u_0, v_0) be an isolated real point of intersection of the loci

$$\frac{1}{2}F_{\xi\xi}u^2 + F_{\xi\varepsilon}u + F_{\eta\varepsilon}v + \frac{1}{2}F_{\varepsilon\varepsilon} = 0,$$

(3.113)

$$\frac{1}{2}G_{\xi\xi}u^2 + G_{\xi\varepsilon}u + G_{\eta\varepsilon}v + \frac{1}{2}G_{\varepsilon\varepsilon} = 0,$$

i.e. an intersection for which the Jacobian

$$\det \begin{Vmatrix} F_{\xi\xi}u_0 + F_{\xi\varepsilon} & F_{\eta\varepsilon} \\ G_{\xi\xi}u_0 + G_{\xi\varepsilon} & G_{\eta\varepsilon} \end{Vmatrix} \tag{3.114}$$

is different from zero. By the standard implicit function theorem, there exists a solution of the system (3.112) for u and v as functions of ε with $u = u_0$ and $v = v_0$ at $\varepsilon = 0$. As a result, the system $F = G = 0$ has a solution for ξ and η as functions of ε, $\xi = \varepsilon u(\varepsilon)$, $\eta = \varepsilon v(\varepsilon)$, with $d\xi/d\varepsilon = u_0$, $d\eta/d\varepsilon = v_0$ at $\varepsilon = 0$. This proves, in case the Jacobian (3.114) is nonzero, the existence of a periodic solution of the type sought.

THEOREM 3.14. *In equation* (1.2) *let* $g(x, x') \in C^2$, *and let* $f(t, x, x', \varepsilon) \in C^1$. *Let equation* (1.1) *have the nonconstant periodic solution* $x_0(t)$ *of least period* L_0. *Let* $f(t, x, x', \varepsilon)$ *have least period* T *in* t, *where* T *is a rational multiple of* L_0. *Let* $x_0(0) = A$, $x_0'(0) = 0$. *Further let all solutions of the variation equation* (3.2) *be periodic, and let*

$$\int_0^L \frac{p(s)}{W(s)} f(s, x_0(s), x_0'(s), 0)\, ds = 0 \tag{3.115}$$

$$\int_0^L \frac{q(s)}{W(s)} f(s, x_0(s), x_0'(s), 0)\, ds = 0. \tag{3.116}$$

Let (u_0, v_0) *be a real solution of the system* (3.113) *for which* (3.114) *is not zero.*

Then there exists a periodic solution of (1.2) *of period* L *of the form*

$$x = x_0(t) + \varepsilon x_1(t) + o(\varepsilon) \tag{3.117}$$

where $x_1(t)$ *is that solution of equation* (3.108) *with* $y(0) = u_0$, $y'(0) = v_0$.

Proof of Theorem 3.14. All of the assertions of the theorem have been proved in the preceding discussion except for the very last. For the periodic solution, $x(0) = A + \xi = A + u_0\varepsilon + o(\varepsilon)$, and $x'(0) = \eta = v_0\varepsilon + o(\varepsilon)$. But from (3.117) we have $x(0) = A + \varepsilon x_1(0) + o(\varepsilon)$, and $x'(0) = \varepsilon x_1'(0) + o(\varepsilon)$. Hence $x_1(0) = u_0$, and $x_1'(0) = v_0$.

Remarks. The Jacobian (3.114) is the critical Jacobian in the theory of Coddington and Levinson and Friedrichs.

In case the loci (3.113) are nondegenerate parabolas, there may be as many as two real points (u_0, v_0), giving two different periodic solutions of the type sought.

In the case that (1.1) has a two-parameter family of periodic solutions of period L_0 (and thus in particular in the quasi-linear case), $F_{\xi\xi} = G_{\xi\xi} = 0$, and the system (3.113) is a linear system of equations whose determinant of coefficients is the Jacobian (3.114).

When conditions (3.115) and (3.116) are not both satisfied, it may be that they will be when a translation $x_0(t + \tau)$ is considered. We shall investigate this later in the present section.

When the Jacobian (3.114) is zero, we can sometimes obtain information from the third derivatives of F and G at $(0, 0, 0)$. We shall do this later in the section.

We note that it is again possible to obtain the initial values of $x_1(t)$ and $x_1'(t)$ by formal manipulation with the perturbation series. The term $x_2(t)$ in the perturbation series will satisfy equation (3.18), where this time $x_1(t)$ is given by

$$x_1(t) = C_1 p(t) + C_1^* q(t) + h_1(t), \tag{3.118}$$

(cf. (3.109)). It is necessary, in order that a periodic $x_2(t)$ exist, that the right member of (3.18) be orthogonal to both $p(t)/W(t)$ and $q(t)/W(t)$ on the interval $(0, L)$. It is found that the orthogonality conditions are equations which determine the constants C_1 and C_1^*. From (3.118), $x_1(0) = C_1^* q(0) = C_1^*/g(A, 0)$, while $x_1'(0) = C_1 p'(0) = -C_1 g(A, 0)$. If we make the substitution $u = C_1^*/g(A, 0)$, $v = -C_1 g(A, 0)$, the two orthogonality conditions give exactly equations (3.113), so that the correct values for C_1 and C_1^* are determined from u_0 and v_0 according to the above substitution, and agree with the results of Theorem 3.14.

We now consider those cases in which the Jacobian (3.114) is zero which can be resolved from a knowledge of the third partial derivatives of F and G at $(0, 0, 0)$. The ten different third partial derivatives of F and the ten different partial derivatives of G at $(0,0,0)$ have expressions similar to those found in earlier sections. Since their explicit expressions are complicated and not of direct interest, we do not give them. Some simplifications can occur in some of them. For example, $F_{\eta\eta\eta}(0,0,0) = G_{\eta\eta\eta}(0,0,0) = 0$. In case the autonomous equation, $x'' + g(x, x') = 0$, has a two-parameter family of periodic solutions of period L_0, all partial derivatives not involving differentiation with respect to ε are zero at $(0,0,0)$.

We assume then, that (u_0, v_0) is a real solution of the system (3.113) for which the Jacobian (3.114) is zero. We shall determine those circumstances under which we may assert the existence of a solution of the system (3.112) for u and v as functions of ε, having $u = u_0$, $v = v_0$ at $\varepsilon = 0$. When this is done, we will know that the system $F = G = 0$ will have a solution for ξ and η as functions of ε with $\xi = \eta = 0$ at $\varepsilon = 0$, and $d\xi/d\varepsilon = u_0$, $d\eta/d\varepsilon = v_0$ at $\varepsilon = 0$. To ensure the existence and continuity at $(0,0,0)$ of the third partial derivatives of F and G, we assume that in the original differential equation, $g(x, x') \in C^3$, and $f(t, x, x', \varepsilon) \in C^2$.

We denote the homogeneous polynomials of degrees 2 and 3 which are at the beginnings of the Taylor series for F and G in powers of ξ, η, and ε by $F_2(\xi, \eta, \varepsilon)$, $F_3(\xi, \eta, \varepsilon)$, $G_2(\xi, \eta, \varepsilon)$, and $G_3(\xi, \eta, \varepsilon)$. The system (3.112) is then

$$(3.119) \qquad \begin{aligned} F_2(u, v, 1) + \varepsilon F_3(u, v, 1) + o(\varepsilon) &= 0, \\ G_2(u, v, 1) + \varepsilon G_3(u, v, 1) + o(\varepsilon) &= 0, \end{aligned}$$

while the system (3.113) is just $F_2(u, v, 1) = G_2(u, v, 1) = 0$.

$$(3.120) \qquad \left\| \begin{matrix} F_{\xi\xi} u_0 + F_{\xi\varepsilon} & F_{\eta\varepsilon} \\ G_{\xi\xi} u_0 + G_{\xi\varepsilon} & G_{\eta\varepsilon} \end{matrix} \right\|$$

must have rank 1. At least one of $F_{\eta\varepsilon}$ and $G_{\eta\varepsilon}$ is nonzero, for if both vanished, the intersection (u_0, v_0) would not be isolated, since the whole line $u = u_0$ would satisfy the system. We may therefore solve one of the equations in (3.119) for v as a function of u and ε near $v = v_0$, $u = u_0$, $\varepsilon = 0$. When the result is substituted into the second equation, we obtain

$$\frac{1}{2}(F_{\xi\xi}G_{\eta\varepsilon} - G_{\xi\xi}F_{\eta\varepsilon})(u - u_0)^2 + (F_3(u_0, v_0, 1)G_{\eta\varepsilon} \tag{3.121}$$
$$- G_3(u_0, v_0, 1)F_{\eta\varepsilon})\varepsilon + \text{higher order terms} = 0.$$

In (3.121) the coefficient of $(u-u_0)^2$ is not zero, since if it were, (u_0, v_0) would not be an isolated solution of the system (3.112). There is no linear term in $(u-u_0)$ in (3.121) because of the vanishing of the Jacobian (3.114).

Now suppose that the coefficient of ε in (3.121) is not zero. If it were zero, the situation could not be resolved without a knowledge of higher order derivatives of F and G. When the coefficient is nonzero, (3.121) can be solved for u as a function of ε, there being two distinct solutions. The solutions exist for ε positive and not for ε negative if

$$\frac{F_3(u_0, v_0, 1)G_{\eta\varepsilon} - G_3(u_0, v_0, 1)F_{\eta\varepsilon}}{F_{\xi\xi}G_{\eta\varepsilon} - G_{\xi\xi}F_{\eta\varepsilon}} \tag{3.122}$$

is negative, and the solutions exist for ε negative and not for ε positive if (3.122) is positive. Hence the system $F = G = 0$ can be solved for ξ and η as a function of ε, there being two solutions for positive ε and none for negative ε if (3.122) is negative, and two solutions for negative ε and none for positive ε if (3.122) is positive.

We now consider the cases in which (u_0, v_0) is a nonisolated solution of the system (3.112). The simplest situation is that in which both $F_2(u, v, 1)$ and $G_2(u, v, 1)$ vanish identically. In this case, the system (3.119) can be simplified to

$$F_3(u, v, 1) + o(1) = 0$$
$$G_3(u, v, 1) + o(1) = 0.$$

Suppose that (u_0, v_0) is a real solution of the system $F_3(u, v, 1) = G_3(u, v, 1) = 0$, for which the Jacobian of F_3 and G_3 with respect to u and v at (u_0, v_0) is not zero. Then we can solve (3.119) for u and v as functions of ε, and hence can solve the system $F = G = 0$ for ξ and η as functions of ε with $\xi = \eta = 0$, $d\xi/d\varepsilon = u_0$, $d\eta/d\varepsilon = v_0$ at $\varepsilon = 0$. There may be as many as nine real solutions in this case.

The next possibility is that in which $F_2(u, v, 1)$ and $G_2(u, v, 1)$ are proportional, without both vanishing identically. (In this case, one of them may still vanish identically.) Let λ and μ be constants, not both zero, such that

$$\lambda F_2(u, v, 1) + \mu G_2(u, v, 1) \equiv 0.$$

Then if, say, $F_2(u, v, 1) \not\equiv 0$, replace the second equation of (3.119) by

$$\varepsilon(\lambda F_3(u, v, 1) + \mu G_3(u, v, 1)) + o(\varepsilon) = 0.$$

Suppose that the system

$$(3.123) \quad \begin{aligned} F_2(u, v, 1) &= 0 \\ \lambda F_3(u, v, 1) + \mu G_3(u, v, 1) &= 0 \end{aligned}$$

has a solution (u_0, v_0) for which the Jacobian with respect to u and v at (u_0, v_0) is not zero. Then the system (3.119) can be solved for u and v, and the system $F = G = 0$ can be solved for ξ and η as functions of ε.

We now turn to the case in which neither $F_2(u, v, 1)$ nor $G_2(u, v, 1)$ vanishes identically, and they are not proportional. If the point (u_0, v_0) is a nonisolated solution of (3.112), the equations must have the forms

$$(3.124) \quad \begin{aligned} (u-u_0)(A_1 + B_1 u) + \varepsilon F_3(u, v, 1) + o(\varepsilon) &= 0 \\ (u-u_0)(A_2 + B_2 u) + \varepsilon G_3(u, v, 1) + o(\varepsilon) &= 0. \end{aligned}$$

where $A_1 B_2 - A_2 B_1 \neq 0$. ($B_1 = \frac{1}{2} F_{\xi\xi}$ and $B_2 = \frac{1}{2} G_{\xi\xi}$.) Since at least one of $(A_1 + B_1 u_0)$ and $(A_2 + B_2 u_0)$ is not zero, one of the equations in (3.124) may be solved for u as a function of ε and v near $u = u_0$, $\varepsilon = 0$. When the result is substituted into the second equation, the resulting equation involving ε and v alone is equivalent to

$$(3.125) \quad \varepsilon((A_1+B_1 u_0) G_3(u_0, v, 1) - (A_2+B_2 u_0) F_3(u_0, v, 1)) + o(\varepsilon) = 0.$$

Now let v_0 be a simple real root of the equation

$$(3.126) \quad H(v) \equiv (A_1+B_1 u_0) G_3(u_0, v, 1) - (A_2+B_2 u_0) F_3(u_0, v, 1) = 0$$

at which, therefore, $H'(v_0) \neq 0$. Then we may solve (3.125) for v as a function of ε, and hence solve the system (3.124) for u and v as functions of ε with $u = u_0$, $v = v_0$ at $\varepsilon = 0$. Thus again we can solve $F = G = 0$ for ξ and η as functions of ε.

The above results exhaust the possibilities, and we summarize them in the following theorem.

THEOREM 3.15. *Let all the hypotheses of Theorem* 3.14 *hold through* (3.116), *and in addition let* $g(x, x') \in C^3$, $f(t, x, x', \varepsilon) \in C^2$. *Let* (u_0, v_0) *be a real solution of the system* (3.113) *for which the Jacobian* (3.114) *is zero. Let one of the following four alternatives be true:*

(a) (u_0, v_0) *is an isolated solution of* (3.113) *and* (3.122) *is not zero.*

(b) $F_2(u, v, 1)$ *and* $G_2(u, v, 1)$ *are not proportional,* (u_0, v_0) *is non-isolated, and* v_0 *is a simple root of* (3.126).

(c) $F_2(u, v, 1)$ *and* $G_2(u, v, 1)$ *are proportional, but not both identically zero and* (u_0, v_0) *is a root with nonvanishing Jacobian of* (3.123) *(or of a correspond-corresponding system if* $F_2 \equiv 0$, $G_2 \neq 0$).

(d) $F_2(u, v, 1)$ *and* $G_2(u, v, 1)$ *both vanish identically, and* (u_0, v_0) *is a root with nonvanishing Jacobian of the system* $F_3(u, v, 1) = G_3(u, v, 1) = 0$.

Then there exist one or two periodic solutions of (1.2) *of period* L *of the form*

$$x = x_0(t) + \varepsilon x_1(t) + o(\varepsilon)$$

where $x_1(t)$ *is that solution of equation* (3.108) *with* $y(0) = u_0$ *and* $y'(0) = v_0$. *There are two solutions in case (a), both existing for positive* ε *and not for negative* ε *if* (3.122) *is negative, and both existing for negative* ε *and not for positive* ε *if* (3.122) *is positive.*

Remarks. When the autonomous equation $x'' + g(x, x') = 0$ has a two-parameter family of periodic solutions of period L_0, only cases (c) and (d) can occur.

In case (b) there may be as many as three points (u_0, v_0), in case (d) as many as nine.

We now consider the stability of the periodic solutions found in Theorems 3.14 and 3.15. The investigation proceeds as in the earlier sections. It is found that

$$\phi(L) = 1 + \varepsilon(u_0 F_{\xi\xi} + F_{\xi\varepsilon}) + o(\varepsilon),$$

$$\phi'(L) = \varepsilon(u_0 G_{\xi\xi} + G_{\xi\varepsilon}) + o(\varepsilon), \tag{3.127}$$

$$\begin{aligned} \psi(L) &= \varepsilon F_{\eta\varepsilon} + o(\varepsilon), \\ \psi'(L) &= 1 + \varepsilon G_{\eta\varepsilon} + o(\varepsilon). \end{aligned} \tag{3.127 (continued)}$$

The two characteristic multipliers, which are the eigenvalues of

$$\begin{Vmatrix} \phi(L) & \psi(L) \\ \phi'(L) & \psi'(L) \end{Vmatrix} \tag{3.128}$$

are given by $\lambda_1 = 1 + \varepsilon\mu_1 + o(\varepsilon)$, $\lambda_2 = 1 + \varepsilon\mu_2 + o(\varepsilon)$, where μ_1 and μ_2 are the eigenvalues of the matrix

$$\begin{Vmatrix} u_0 F_{\xi\xi} + F_{\xi\varepsilon} & F_{\eta\varepsilon} \\ u_0 G_{\xi\xi} + G_{\xi\varepsilon} & G_{\eta\varepsilon} \end{Vmatrix} \tag{3.129}$$

Now the matrix (3.129) is the matrix of the Jacobian (3.114), so that neither μ_1 nor μ_2 is zero under the hypothesis of Theorem 3.14. As a result, under the hypothesis of Theorem 3.14, both characteristic multipliers differ from +1, and we can analyze the stability.

If the Jacobian (3.114) is negative, the two μ's are real and of opposite sign, so that the λ's are real and separated by 1 for small ε. Thus the periodic solution is directly unstable in this case.

If the Jacobian (3.114) is positive, the two μ's are real with the same sign, or else conjugate complex. Their real parts have the same sign, determined by the trace

$$u_0 F_{\xi\xi} + F_{\xi\varepsilon} + G_{\eta\varepsilon}.$$

Referring to (3.109), we see that the real parts of the λ's will be greater or less than 1 for small ε according as

$$\begin{aligned} \varepsilon \int_0^L [& - g_{xx'}(x_0(s), x_0'(s))(u_0 g(A, 0) q(s) + h_1(s)) \\ & - g_{x'x'}(x_0(s), x_0'(s))(u_0 g(A, 0) g'(s) + h_1'(s)) \\ & + f_{x'}(s, x_0(s), x_0'(s), 0)] ds \end{aligned} \tag{3.130}$$

is positive or negative.

THEOREM 3.16. *Let all of the hypotheses of Theorem* 3.14 *be satisfied. Then the periodic solution found in that theorem has the following stability properties:*

(a) *If the Jacobian* (3.114) *is negative, the solution is directly unstable.*

(b) *If the Jacobian* (3.114) *is positive, the solution is completely stable of completely unstable according as* (3.130) *is negative or positive.*

Proof of Theorem 3.16: The proof is contained in the discussion preceding the theorem.

Remarks. When $u_0 F_{\xi\xi} = 0$, the expression (3.130) is the same as (3.90) in Theorem 3.9.

When (3.114) is positive and (3.130) vanishes, the distinction can still be made between complete stability and complete instability in certain cases, which will be investigated later in this section.

We now investigate the stability of the periodic solutions found in Theorem 3.15. It will be found that the quantity which distinguishes between direct instability on the one hand and complete stability or complete instability on the other hand is in each of the cases of Theorem 3.15, the quantity assumed not to vanish. The quantity which distinguishes between complete stability and complete instability will again be (3.130) when (3.130) does not vanish.

Of the four cases of Theorem (3.15), case (a) is distinguished because it involves fractional powers of ε whereas the remaining cases do not. In case (a) there occur two determinants, which we denote by B_1 and B_2:

$$
\begin{aligned}
B_1 &= F_{\xi\xi} G_{\eta\varepsilon} - G_{\xi\xi} F_{\eta\varepsilon} \\
B_2 &= F_3(u_0, v_0, 1) G_{\eta\varepsilon} - G_3(u_0, v_0, 1) F_{\eta\varepsilon} .
\end{aligned}
\tag{3.131}
$$

From (3.121) it follows that

$$u = u_0 \pm \sqrt{-\frac{2B_2}{B_1} \varepsilon \operatorname{sgn} \varepsilon} + o(\sqrt{|\varepsilon|})$$

$$\xi = \varepsilon u = \varepsilon u_0 \pm \operatorname{sgn} \varepsilon |\varepsilon|^{3/2} \sqrt{-\frac{2B_2}{B_1} \operatorname{sgn} \varepsilon} + o(|\varepsilon|^{3/2}).$$

By calculations similar to those in earlier sections we find that in case (a)

$$\phi(L) = 1 + \varepsilon(F_{\xi\xi}u_0 + F_{\xi\varepsilon} \pm F_{\xi\xi}\ \mathrm{sgn}\ \varepsilon|\varepsilon|^{3/2}\sqrt{-\frac{2B_2}{B_1}\ \mathrm{sgn}\ \varepsilon} + o(|\varepsilon|^{3/2})$$

$$\phi'(L) = \varepsilon(G_{\xi\xi}u_0 + G_{\xi\varepsilon}) \pm G_{\xi\xi}\ \mathrm{sgn}\ \varepsilon|\varepsilon|^{3/2}\sqrt{-\frac{2B_2}{B_1}\ \mathrm{sgn}\ \varepsilon} + o(|\varepsilon|^{3/2})$$

$$\psi(L) = \varepsilon F_{\eta\varepsilon} + o(|\varepsilon|^{3/2})$$

$$\psi'(L) = 1 + \varepsilon G_{\eta\varepsilon} + o(|\varepsilon|^{3/2}).$$

Using the fact that the Jacobian (3. 114) is zero, we find that $(\lambda_1 - 1)(\lambda_2 - 1)$, where λ_1 and λ_2 are the characteristic multipliers, is given by

$$(3.132)\quad \begin{aligned}(\lambda_1 - 1)(\lambda_2 - 1) &= \pm|\varepsilon|^{5/2} B_1 \sqrt{-\frac{2B_2}{B_1}\ \mathrm{sgn}\ \varepsilon} + o(|\varepsilon|^{5/2}) \\ &= \pm|\varepsilon|^{5/2}\ \mathrm{sgn}\ B_1 \sqrt{-2B_1B_2\ \mathrm{sgn}\ \varepsilon} + o(|\varepsilon|^{5/2}).\end{aligned}$$

Hence for one choice of sign, $(\lambda_1 - 1)(\lambda_2 - 1)$ is positive, and for the other choice it is negative, so that of the two periodic solutions which occur in case (a), exactly one is directly unstable, the other being either completely stable or completely unstable.

The periodic solution last mentioned is completely stable or completely unstable according as (3. 130) is negative or positive, whenever (3. 130) does not vanish. If (3. 130) vanishes, the distinguishing quantity is

$$(3.133)\quad \pm F_{\xi\xi}\ \mathrm{sgn}\ \varepsilon\ |\varepsilon|^{3/2}\sqrt{-\frac{2B_2}{B_1}\ \mathrm{sgn}\ \varepsilon},$$

and it follows from (3. 132) and (3. 133) that there will be complete stability or complete instability according as $F_{\xi\xi}\mathrm{sgn}\,(B_1\varepsilon)$ is negative or positive.

In cases (b), (c), and (d) of Theorem 3. 15 there are no fractional powers of ε. It is found that

$$(3.134)\quad \begin{aligned}\phi(L) &= 1 + \varepsilon(F_{\xi\xi}u_0 + F_{\xi\varepsilon}) + \varepsilon^2(F_{\xi\xi}u_1 + \frac{\partial F_3}{\partial\xi}(u_0, v_0, 1)) + o(\varepsilon^2), \\ \phi'(L) &= \varepsilon(G_{\xi\xi}u_0 + G_{\xi\varepsilon}) + \varepsilon^2(G_{\xi\xi}u_1 + \frac{G_3}{\partial\xi}(u_0, v_0, 1)) + o(\varepsilon^2),\end{aligned}$$

(3.134)
(continued)

$$\psi(L) = \varepsilon F_{\eta\varepsilon} + \varepsilon^2 \frac{\partial F_3}{\partial \eta}(u_0, v_0, 1) + o(\varepsilon^2)$$

$$\psi'(L) = 1 + \varepsilon G_{\eta\varepsilon} + \varepsilon^2 \frac{\partial G_3}{\partial \eta}(u_0, v_0, 1) + o(\varepsilon^2).$$

In (3.134), $u_1 = x_2(0)$, x_2 being the term in the perturbation series $x = x_0(t) + \varepsilon x_1(t) + \varepsilon^2 x_2(t) + \dots$. It will be found that u_1 drops out of most of the expressions which determine stability. In fact, the expressions in which it does not drop out cannot be used in our present theory, because $x_2(t)$ is not completely determined in the situation of Theorem 3.15.

A remark should be made about the partial derivatives occurring in (3.134). The functions F_3 and G_3 were defined above (3.119) as functions of (ξ, η, ε). However, the partial derivatives in (3.134) could also be written

$$\frac{\partial F_3}{\partial \xi}(u_0, v_0, 1) = \frac{\partial}{\partial u_0} F_3(u_0, v_0, 1)$$

etc. We do not use this latter notation because u_0 and v_0 are constants.

The product $(\lambda_1 - 1)(\lambda_2 - 1)$ is given in cases (b) and (c), where not all four of $F_{\xi\xi}u_0 + F_{\xi\varepsilon}$, $G_{\xi\xi}u_0 + G_{\xi\varepsilon}$, $G_{\eta\varepsilon}$, and $G_{\eta\varepsilon}$ vanish, by

(3.135)

$$\begin{aligned}(\lambda_1 - 1)(\lambda_2 - 1) = \varepsilon^3 \Big(B_1 u_1 &+ (F_{\xi\xi}u_0 + F_{\xi\varepsilon}) \frac{\partial G_3}{\partial \eta}(u_0, v_0, 1) \\ &- (G_{\xi\xi}u_0 + G_{\xi\varepsilon}) \frac{\partial F_3}{\partial \eta}(u_0, v_0, 1) - F_{\eta\varepsilon} \frac{\partial G_3}{\partial \xi}(u_0, v_0, 1) \\ &+ G_{\eta\varepsilon} \frac{\partial F_3}{\partial \xi}(u_0, v_0, 1)\Big) + o(\varepsilon^3),\end{aligned}$$

while in case (d), where all the above-mentioned quantities do vanish,

(3.136)
$$(\lambda - 1)(\lambda - 1) = \varepsilon^4 \left[\frac{\partial F_3}{\partial \xi} \frac{\partial G_3}{\partial \eta} - \frac{\partial F_3}{\partial \eta} \frac{\partial G_3}{\partial \xi} \right] + o(\varepsilon^4)$$

all partial derivatives in (3.136) being evaluated at $(u_0, v_0, 1)$.

In both cases (b) and (c), the determinant B_1 is zero, so that the term involving u_1 drops out. There is a possibility in case (c) that all four of the quantities mentioned might be zero (where both $F_2(u, v, 1)$ and $G_2(u, v, 1)$ are constant multiples of $(u - u_0)^2$), but the critical Jacobian always vanishes in this case, so it is not included in case (c).

In case (b), both of $F_{\eta\varepsilon}$ and $G_{\eta\varepsilon}$ are zero, so (3.135) simplifies somewhat. Referring to the function $H(v)$ in (3.126), we find that (3.135) reduces in this case to

$$(3.137) \qquad (\lambda_1 - 1)(\lambda_2 - 1) = \varepsilon^3 H'(v_0) + o(\varepsilon^3),$$

so that stability is determined by the sign of $\varepsilon H'(v_0)$. The second quantity used in stability is again (3.130) provided it does not vanish. If (3.130) vanishes, the distinguishing quantity is

$$(3.138) \qquad F_{\xi\xi} u_1 + \frac{\partial F_3}{\partial \xi}(u_0, v_0, 1) + \frac{\partial G_3}{\partial \eta}(u_0, v_0, 1),$$

but this can be used only if $F_{\xi\xi} = 0$, since u_1 cannot be determined in the situation of Theorem 3.15.

In case (c), (3.135) is found to involve the Jacobian of the system (3.123) multiplied by ε^3/μ (or by $-\varepsilon^3/\lambda$ in case the first equation of (3.123) is $G_2(u, v, 1) = 0$). The second distinguishing quantity is (3.130) or (3.138), with the proviso again that $F_{\xi\xi} = 0$ when (3.138) is used.

In case (d), (3.136) involves exactly the Jacobian of the system $F_3(u, v, 1) = G_3(u, v, 1) = 0$. The second quantity is (3.138), since (3.130) is automatically zero, and moreover in (3.138) $F_{\xi\xi} = 0$ in case (d) automatically.

We now summarize all the above results in a theorem.

THEOREM 3.17. *Let all of the hypotheses of Theorem* 3.15 *hold. Then the periodic solutions found in that theorem have the following stability properties:*

(a) *The periodic solutions in case (a) include one which is directly unstable. The other is completely stable or completely unstable according as* (3.130) *is negative or positive, or in case* (3.130) *vanishes, according as* $F_{\xi\xi}B_1$ *is negative or positive.*

(b) *The periodic solution found in case (b) is directly unstable if* $\varepsilon H'(v_0)$ *is negative. If* $\varepsilon H'(v_0)$ *is positive, the solution is completely stable or completely unstable according as* (3.130) *is negative or positive, or in case* (3.130) *vanishes, and* $F_{\xi\xi}$ *vanishes, according as*

$$\frac{\partial F_3}{\partial \xi}(u_0, v_0, 1) + \frac{\partial G_3}{\partial \eta}(u_0, v_0, 1) \tag{3.139}$$

is negative or positive.

(c) *The periodic solution found in case (c) is directly unstable if*

$$\varepsilon[(F_{\xi\xi}u_0 + F_{\xi\varepsilon})\frac{\partial G_3}{\partial \eta}(u_0, v_0, 1) - (G_{\xi\xi}u_0 + G_{\xi\varepsilon})\frac{\partial F_3}{\partial \eta}(u_0, v_0, 1) - F_{\eta\varepsilon}\frac{\partial G_3}{\partial \xi}(u_0, v_0, 1) + G_{\eta\varepsilon}\frac{\partial F_3}{\partial \xi}(u_0, v_0, 1)] \tag{3.140}$$

is negative. If (3.140) is positive, the solution is completely stable or completely unstable according as (3.130) is negative or positive, or in case vanishes, and $F_{\xi\xi}$ *vanishes, according as (3.139) is negative or positive.*

(d) *The periodic solution found in case (d) is directly unstable if*

$$\frac{\partial F_3}{\partial \xi}\frac{\partial G_3}{\partial \eta} - \frac{\partial F_3}{\partial \eta}\frac{\partial G_3}{\partial \xi}(u_0, v_0, 1) \tag{3.141}$$

is negative. If (3.141) is positive, the solution is completely stable or completely unstable according as (3.139) is negative or positive.

Proof of Theorem 3.17: The largest part of the proof is contained in the discussion which precedes it. We should remark that (3.139) is (3.138) with its first term omitted. In case (c), the quantity (3.140) never vanishes, since it is a non-zero multiple of the critical Jacobian in case (c) of Theorem 3.15. A similar remark holds for (3.141).

Remark: If in the situation of Theorem 3.16, (3.130) vanishes, then (3.138) may be used in place of (3.130), provided that $g(x, x, ') \in C^3$ and $f(t, x, x', \varepsilon) \in C^2$. In the situation of Theorem 3.16, the term $x_2(t)$ can be computed, and u_1 thus determined.

In Sections 3.2 and 3.3 we examined cases in which there were periodic solutions near to translations, $x_0(t + \tau)$ of $x_0(t)$. We shall make a similar investigation in the present section.

When the unperturbed equation, $x'' + g(x, x') = 0$ has a one-parameter family of periodic solutions of period L_0, namely $x_0(t)$ and its translations, it is usually not to be expected that any periodic solutions will be predicted by the results of this section near to translations of $x_0(t)$. The reason for this is that conditions (3.107) involve two requirements, and we have but one parameter to vary. Indeed, when $f(t, x, x', \varepsilon)$ is replaced by $f(t - \tau, x, x', \varepsilon)$, conditions (3.107) become

$$\int_0^L \frac{p(s)}{W(s)} f(s - \tau, x_0(s), x_0'(s), 0)\, ds = 0,$$

$$\int_0^L \frac{q(s)}{W(s)} f(s - \tau, x_0(s), x_0'(s), 0)\, ds = 0.$$

It will usually be impossible to find a single value of τ which will satisfy both these equations.

However, when we consider the case that $x'' + g(x, x')$ has a two-parameter family of periodic solutions of period L_0, we have enough extra freedom to allow a theory to be constructed, analogous to that in Sections 3.2 and 3.3. Suppose that for an interval of values of A, the solution of $x'' + g(x, x') = 0$ with $x(0) = A$, $x'(0) = 0$ is periodic of least period L_0. Let us denote this solution by $x_0(t, A)$. We hope to find A and τ such that there is a periodic solution near to $x_0(t+\tau, A)$.

As before we take the translation τ into account by changing the forcing term to $f(t - \tau, x, x', \varepsilon)$. The variation equation associated with $x_0(t, A)$ is

$$y'' + g_{x'}(x_0(t, A), x_0'(t, A))y' + g_x(x_0(t, A), x_0'(t, A))y = 0. \tag{3.142}$$

We define solutions $p(t, A)$ and $q(t, A)$ by the initial conditions $p(0, A) = 0$, $p'(0, A) = -g(A, 0)$, $q(0, A) = 1/g(A, 0)$, $q'(0, A) = 0$. Both $p(t, A)$ and $q(t, A)$ are periodic, and $p(t, A) \equiv x_0'(t, A)$. The Wronskian $W(t, A) = p(t, A)q'(t, A) - p'(t, A)q(t, A)$ satisfies $W(0, A) = 1$, and is periodic of period L_0. The conditions (3.107) for existence of a periodic solution can be written

$$F_1(A, \tau) \equiv \frac{1}{g(A, 0)} \int_0^L \frac{p(s, A)}{W(s, A)} f(s - \tau, x_0(s, A), x_0'(s, A), 0)\, ds = 0 \tag{3.143}$$

$$G_1(A, \tau) \equiv g(A, 0) \int_0^L \frac{q(s, A)}{W(s, A)} f(s - \tau, x_0(s, A), x_0'(s, A), 0)\, ds = 0$$

If we let $x(t, \xi, \eta, \varepsilon, \tau)$ be that solution of

$$x'' + g(x, x') = \varepsilon f(t - \tau, x, x', \varepsilon) \tag{3.144}$$

having $x = \xi$, $x' = \eta$ at $t = 0$, and define the functions $F(\xi, \eta, \varepsilon, \tau)$ and $G(\xi, \eta, \varepsilon, \tau)$ by

$$F(\xi, \eta, \varepsilon, \tau) = x(L, \xi, \eta, \varepsilon, \tau) - \xi$$

$$G(\xi, \eta, \varepsilon, \tau) = x'(L, \xi, \eta, \varepsilon, \tau) - \eta,$$

we have that $x_0(t, A) \equiv x(t, A, 0, 0, \tau)$, and that we are attempting to solve $F = G = 0$ near the point $\xi = A$, $\eta = 0$, $\varepsilon = 0$. Since $F(\xi, \eta, 0, \tau) \equiv G(\xi, \eta, 0, \tau) \equiv 0$, all partial derivatives of F and G at $\eta = \varepsilon = 0$, not involving differentiation with respect to ε, will vanish. It follows from the results stated earlier in the section that

$$F_1(A, \tau) = F_\varepsilon(A, 0, 0, \tau), \qquad G_1(A, \tau) = G_\varepsilon(A, 0, 0, \tau).$$

The partial derivatives of $F_1(A, \tau)$ and $G_1(A, \tau)$ with respect to A are derivatives of F_ε and G_ε with respect to ξ,

$$\frac{\partial F_1}{\partial A} = F_{\xi\varepsilon}(A, 0, 0, \tau) \qquad \frac{\partial G_1}{\partial A} = G_{\xi\varepsilon}(A, 0, 0, \tau).$$

The partial derivatives with respect to τ can be evaluated by reference to (3.109). The results are

$$\frac{\partial F_1}{\partial \tau} = -\frac{1}{g(A, 0)} F_{\eta\varepsilon}(A, 0, 0, \tau), \qquad \frac{\partial G_1}{\partial \tau} = -\frac{1}{g(A, 0)} G_{\eta\varepsilon}(A, 0, 0, \tau).$$

Therefore the Jacobian (3.114) is $-g(A, 0)$ times the Jacobian of F_1 and G_1 with respect to A and τ, since the derivatives $F_{\xi\xi}$ and $G_{\xi\xi}$ vanish.

THEOREM 3.18. *Let all of the hypotheses of Theorem* 3.14 *preceding* (3.115) *be satisfied. Let* $F_1(A, \tau) = G_1(A, \tau) = 0$, *and let their Jacobian be nonzero. Then there exists for small nonzero* ε *a periodic solution of*

$$x'' + g(x, x') = \varepsilon f(t - \tau, x, x', \varepsilon) \tag{3.144}$$

of period L, *having the form*

$$x = x_0(t, A) + \varepsilon x_1(t) + o(\varepsilon), \tag{3.145}$$

and correspondingly a solution of

(3.146) $$x'' + g(x, x') = \varepsilon f(t, x, x', \varepsilon)$$

of period L *having the form*

(3.147) $$x = x_0(t + \tau, A) + \varepsilon x_1(t + \tau) + o(\varepsilon).$$

If the quantity

(3.148) $$-g(A, 0)\left(\frac{\partial F_1}{\partial A}\frac{\partial G_1}{\partial \tau} - \frac{\partial F_1}{\partial \tau}\frac{\partial G_1}{\partial A}\right)$$

is negative, the solution is directly unstable. If (3.148) *is positive, the solution is completely stable or completely unstable according as*

(3.149) $$\varepsilon\left(\frac{\partial F_1}{\partial A} - g(A, 0)\frac{\partial G_1}{\partial \tau}\right)$$

is negative or positive.

Proof of Theorem 3.18. The theorem is a restatement in the present notation of Theorems 3.14 and 3.16.

Remark: A similar theorem can be obtained in a case corresponding to Theorems 3.15 and 3.17. Consider the case in which $F_1(A, \tau) \equiv G_1(A, \tau) \equiv 0$. All second derivatives of F and G except possibly $F_{\varepsilon\varepsilon}$ and $G_{\varepsilon\varepsilon}$ now vanish, and all third derivatives not involving at least two differentiations with respect to ε vanish. Define functions $F_2(A, \tau)$ and $G_2(A, \tau)$ in terms of $F_{\varepsilon\varepsilon}$ and $G_{\varepsilon\varepsilon}$. Then the periodic solution will exist if F_2 and G_2 vanish but their Jacobian does not. This corresponds to case (d) of Theorem 3.1. Stability is similarly determined in terms of the eigenvalues of the Jacobian matrix.

Appendix on the Quasilinear Case

A very important case included under Case IV is the so-called quasilinear case, in which the unperturbed equation is linear, i.e. $g(x, x') \equiv -4\pi^2 x/L_0^2$. In this case, many of the partial derivatives of F and G have very simple expressions, and in this appendix we shall list a few cases.

Because there is a two-parameter family of periodic solutions of $x'' + g(x, x') = 0$, (indeed all solutions are periodic), the partial derivatives not involving differentiation with respect to ε will all vanish automatically. The solution $x_0(t, A)$ is, of course, just $A \cos \frac{2\pi t}{L_0}$. The various partial derivatives which will be evaluated turn out to be Fourier coefficients of the functions

$$f(t, x_0(t), x_0'(t), 0), \qquad f_x(t, x_0(t), x_0'(t), 0), \qquad f_{x'}(t, x_0(t), x_0'(t), 0).$$

The Fourier Series for the first of these will be

$$\frac{1}{2}a_0 + \sum_{n=1}^{\infty} a_n \cos \frac{2n\pi t}{L} + b_n \sin \frac{2n\pi t}{L}, \tag{3.150}$$

while the coefficients for the remaining two functions will be denoted by a_{nx}, b_{nx}, and $a_{nx'}$, $b_{nx'}$ respectively.

Let p and q be relatively prime integers such that $L = pT = qL_0$. Then we have the following formulas for partial derivatives:

$$(3.151) \quad \begin{aligned}
F_\varepsilon &= -\frac{qL_0^2}{4\pi} a_q \\
G_\varepsilon &= \frac{qL_0}{2} b_q \\
F_{\xi\varepsilon} &= \frac{qL_0}{4} (a_{ox'} - a_{2qx'}) - \frac{qL_0^2}{8\pi} b_{2qx} \\
G_{\eta\varepsilon} &= \frac{qL_0}{4} (a_{ox} + a_{2qx}) - \frac{\pi q}{2} b_{2qx'} \\
F_{\xi\varepsilon} &= -\frac{qL_0^3}{16\pi^2} (a_{ox} - a_{2qx}) - \frac{qL_0^2}{8\pi} b_{2qx'} \\
G_{\eta\varepsilon} &= \frac{qL_0}{4} (a_{ox'} + a_{2qx'}) + \frac{qL_0^2}{8\pi} b_{2qx}
\end{aligned}$$

The requirement of Theorem 3.14 is then that $a_q = b_q = 0$, while

$$(3.152) \quad \begin{aligned}
&\frac{q^2 L_0^2}{16} (a_{ox'}^2 - a_{2qx'}^2 - b_{2qx'}^2) + \frac{q^2 L_0^3}{16\pi} (a_{2qx'} b_{2qx'} - a_{2qx'} b_{2qx}) \\
&\quad + \frac{q^2 L_0^4}{64\pi^2} (a_{ox}^2 - a_{2qx}^2 - b_{2qx}^2)
\end{aligned}$$

should be different from zero. The periodic solution is directly unstable when (3.152) is negative, and when (3.152) is positive, the solution is completely stable or completely unstable according as $\varepsilon q L_0 a_{ox'}/2$ is negative or positive.

As an example of this last, consider van der Pol's equation with a small parameter,

$$x'' + x = \varepsilon(1 - x^2)x' + \varepsilon E \cos(t + \phi). \tag{3.153}$$

In (3.153), $L_0 = T = 2\pi$, so $p = q = 1$. It is readily found that

$$a_1 = E \cos \phi, \qquad b_1 = \frac{A^3}{4} - A - E \sin \phi,$$

$$a_3 = \frac{A^3}{4}, \qquad b_{2x} = A^2, \qquad a_{ox'} = 2 - A^2,$$

$$a_{2x'} = -\frac{A^2}{2}.$$

All remaining coefficients are zero. In order that $a_1 = b_1 = 0$, $\cos \phi = 0$, and A must be a root of $\frac{A^3}{4} - A = \pm E$. The determinant (3.152) becomes

$$\frac{\pi^2}{16}(4 - 3A^2)(4 - A^2),$$

with the second distinguishing quantity equal to $\varepsilon\pi(2 - A^2)$.

It follows from the above that if $0 < |E| < \frac{4\sqrt{3}}{9}$, there are three periodic solutions of (3.153), having A^2 in the ranges $0 < A^2 < 4/3$, $4/3 < A^2 < 4$, and $4 < A^2$. When ε is positive, the three solutions are completely unstable, directly unstable, and completely stable, respectively. If $|E| > \frac{4\sqrt{3}}{9}$, there is only one periodic solution, having $A^2 > 4$, completely stable for ε positive. When ε is negative, complete stability and complete instability replace one another. When $E = 0$, or when $E = \pm\frac{4\sqrt{3}}{9}$, the critical Jacobian vanishes for the two solutions which then have $A^2 = 4/3$. This marks a coalescence of two periodic solutions as E passes through the critical value. The third, for which $A^2 = 16/3$, exists by the theorem and is completely stable if ε is positive.

PART IV: AUTONOMOUS PERTURBING TERMS

4.1. Introduction

In this part of the paper we consider the case of the autonomous equation

$$x'' + g(x, x') = \varepsilon f(x, x', \varepsilon). \tag{4.1}$$

The periodic solution $x = x_0(t)$ of the unperturbed equation

$$x'' + g(x, x') = 0 \tag{4.2}$$

is assumed to be nonconstant and of least period L_0. [For the case that $x_0(t)$ is constant, see Part II.].

It is well known, cf. [3] Chapt. 13, that under certain conditions (4.1) will have a periodic solution near to $x_0(t)$. The period will in general differ from the period of $x_0(t)$. Using the methods of the earlier parts of this paper, we are able to find expressions for such periodic solutions.

We find, as in Part III, that Case I cannot occur. Case II turns out to be particularly simple. The real difficulties arise in Cases III and IV.

The details of the investigation differ slightly from Parts II and III. Since equation (4.1) is autonomous, if $x(t)$ is a periodic solution of (4.1), any translation $x(t + \tau)$ is also a periodic solution. Any technique of finding a periodic solution $x(t)$ near to $x_0(t)$ for small ε would inevitably find all translations $x(t + \tau)$ for small τ. To avoid this ambiguity we agree to impose the further condition $x'(0) = 0$. This is in accordance with the fundamental assumption that $x_0'(0) = 0$.

We assume as before that $g(x, x') \in C^2$, $f(x, x', \varepsilon) \in C^1$. Let $x_0(0) = A$, $x_0'(0) = 0$. Then $g(A, 0) \neq 0$, for if $g(A, 0) = 0$, we would have $x_0(t) \equiv A$, contrary to the assumption that $x_0(t)$ is nonconstant. Let $x(t, \xi, \varepsilon)$ be that solution of (4.1) for which $x(0, \xi, \varepsilon) = A + \xi$, $x'(0, \xi, \varepsilon) = 0$. Note that $x(t, 0, 0,) \equiv x_0(t)$. We shall find, for small ε, such values of ξ as make $x(t, \xi, \varepsilon)$ periodic.

The period of $x(t, \xi, \varepsilon)$, when it is periodic, will not be L_0 in general, but will differ slightly from L_0. Define $F(\tau, \xi, \varepsilon)$ and $G(\tau, \xi, \varepsilon)$ by

$$F(\tau, \xi, \varepsilon) = x(L_0 + \tau, \xi, \varepsilon) - A - \xi,$$
(4.3)
$$G(\tau, \xi, \varepsilon) = x'(L_0 + \tau, \xi, \varepsilon).$$

Then $F(0, 0, 0) = G(0, 0, 0) = 0$, and if $F(\tau, \xi, \varepsilon) = G(\tau, \xi, \varepsilon) = 0$, the solution $x(t, \xi, \varepsilon)$ will be periodic with period $L_0 + \tau$. We shall solve the system

(4.4) $$F(\tau, \xi, \varepsilon) = G(\tau, \xi, \varepsilon) = 0$$

for τ and ξ as functions of ε for small ε in the sense that existence of the solutions will be established, and the derivatives $d\tau/d\varepsilon$ and $d\xi/d\varepsilon$ will be evaluated at $\varepsilon = 0$.

We will thus have proved the existence of a periodic solution of (4.1) near to $x_0(t)$, and will have an approximation to its initial value, $A + \xi$. If we make the change of independent variable

(4.5) $$s = \frac{L_0 t}{L_0 + \tau(\varepsilon)},$$

in (4.1), we shall obtain a differential equation whose solutions are periodic functions of s, all with period L_0. The new differential equation is

$$\frac{d^2x}{ds^2} + g(x, \frac{dx}{ds}) = \varepsilon[f(x, \frac{dx}{ds}, 0) - \frac{2\tau'(0)}{L_0} g(x, \frac{dx}{ds}) + \frac{\tau'(0)}{L_0} g_{x'}(x, \frac{dx}{ds}) \frac{dx}{ds}] + o(\varepsilon)$$
(4.6)

The periodic solution of (4.6) is

(4.7) $$x = x_0(s) + \varepsilon x_1(s) + o(\varepsilon).$$

where $x_1(s)$ is that periodic solution of

$$y'' + g_{x'}(x_0(s), x_0'(s))y' + g_x(x_0(s), x_0'(s))y$$

(4.8) $$= f(x_0(s), x_0'(s), 0) - \frac{2\tau'(0)}{L_0} g(x_0(s), x_0'(s)$$

$$+ \frac{\tau'(0)}{L_0} g_{x'}(x_0(s), x_0'(s))x_0'(s),$$

which satisfies $x_1'(0) = 0$. On returning to t as independent variable we find

(4.9) $$x = x_0\left(\frac{L_0 t}{L_0 + \tau(\varepsilon)}\right) + \varepsilon x_1\left(\frac{L_0 t}{L_0 + \tau(\varepsilon)}\right) + o(\varepsilon).$$

In the more involved situations which will occur in Cases III and IV we shall need more complete information on the form of (4.6).

The question of stability of a periodic solution of an autonomous system involves the concept of orbital stability. The variation equation of (4.1) with respect to the periodic solution (4.9)

$$y'' + g_x(x, x')y + g_x(x, x')y = \varepsilon f_x(x, x', \varepsilon) + \varepsilon f_{x'}(x, x', \varepsilon) \tag{4.10}$$

or the variation equation of (4.6) with respect to the periodic solution (4.7)

$$\begin{aligned} y'' + g_{x'}(x, x')y' + g_x(x, x')y = \varepsilon \{ & f_x(x, x', 0)y \\ & + f_{x'}(x, x', 0)y' - \frac{2\tau'(0)}{L_0}[g_x(x, x')y + g_{x'}(x, x')y'] \\ & + \frac{\tau'(0)}{L_0}[g_x(x, x')y' + g_{xx'}(x, x')x'y + g_{x'x'}(x, x')x'y']\} \\ & + o(\varepsilon) \end{aligned} \tag{4.11}$$

has one of its characteristic multipliers equal to +1. If the second characteristic multiplier is less than 1 in absolute value, the periodic solution (4.9) or (4.7) is orbitally stable [3] Chapt. 14. This means that if $x(t)$ is the periodic solution, any solution which at $t = 0$ is sufficiently close to the orbit $x = x(t)$, $x' = x'(t)$ will converge, as $t \to +\infty$, to some translation of $x(t)$. Thus the question of stability involves knowing for small ε the absolute value of the second characteristic multiplier of the variational equation.

4.2. Existence and Stability in Case II

We use the same notation in this section as in Section 3.2. In particular, the fundamental set of solutions of the variation equation

$$y'' + g_{x'}(x_0(t), x_0'(t))y' + g_x(x_0(t), x_0'(t))y = 0 \tag{4.12}$$

is $p(t) \equiv x_0'(t)$, and $q(t)$, where $q(0) = 1/g(A, 0)$, and $q(t)$ has the form $q(t) = r(t)e^{-\alpha t}$, with $r(t)$ periodic and α given by

$$\alpha = \frac{1}{L_0}\int_0^{L_0} g_{x'}(x_0(s), x_0'(s))\,ds \neq 0. \tag{4.13}$$

The characteristic multipliers of (4.11) are 1 and $e^{-\alpha L_0}$. The Wronskian $W(t)$ of $p(t)$ and $q(t)$ is given by

$$(4.14) \qquad W(t) \equiv p(t)q'(t) - q(t)p'(t) = e^{-\int_0^t g_{x'}(x_0(s), x_0'(s))\,ds},$$

so that $W(0) = 1$, and $W(t)$ is a periodic function multiplied by $e^{-\alpha t}$.

To evaluate the first partial derivatives of $F(\tau, \xi, \varepsilon)$ and $G(\tau, \xi, \varepsilon)$ (cf. (4.3)) at $(0, 0, 0)$, we need the partial derivatives $x_\xi(t, \xi, \varepsilon)$ and $x_\varepsilon(t, \xi, \varepsilon)$ for $\xi = \varepsilon = 0$. We find

$$(4.15) \qquad x_\xi(t, 0, 0) = \varphi(t) = g(A, 0)q(t) + q'(0)p(t).$$

$$(4.16) \qquad x_\varepsilon(t, 0, 0) = \int_0^t \frac{p(s)q(t) - q(s)p(t)}{W(s)} f(x_0(s), x_0'(s), 0)\,ds.$$

Using these we find that

$$F_\tau(0,0,0) = x'(L_0, 0, 0) = x_0'(L_0) = 0,$$

$$G_\tau(0,0,0) = x''(L_0, 0, 0) = x_0''(L_0) = -g(A, 0) \neq 0,$$

$$F_\xi(0,0,0) = x_\xi(L_0, 0, 0) - 1 = e^{-\alpha L_0} - 1 \neq 0,$$

$$(4.17) \qquad G_\xi(0,0,0) = x_\xi'(L_0, 0, 0) = g(A, 0)q'(0)(e^{-\alpha L_0} - 1)$$

$$F_\varepsilon(0,0,0) = x_\varepsilon(L_0, 0, 0) = \frac{e^{-\alpha L_0}}{g(A, 0)} \int_0^{L_0} \frac{p(s)}{W(s)} f(x_0(s), x_0'(s), 0)\,ds,$$

$$G_\varepsilon(0,0,0) = x_\varepsilon'(L_0, 0, 0) = e^{-\alpha L_0} q'(0) \int_0^{L_0} \frac{p(s)}{W(s)} f(x_0(s), x_0'(s), 0)\,ds$$

$$+ g(A, 0) \int_0^{L_0} \frac{q(s)}{W(s)} f(x_0(s), x_0'(s), 0)\,ds.$$

From (4.17) we see that the Jacobian of F and G with respect to τ and ξ at $(0, 0, 0)$ has the value

$$g(A, 0)(e^{-\alpha L_0} - 1) \neq 0.$$

Therefore in Case II we may always solve for τ and ξ uniquely as functions of ε for small ε. The derivatives of τ and ξ with respect to ε at $\varepsilon = 0$ are given by

$$\frac{d\tau}{d\varepsilon} = \int_0^{L_0} \frac{q(s)}{W(s)} f(x_0(s), x_0'(s), 0)\, ds,$$

(4.18)

$$\frac{d\xi}{d\varepsilon} = \frac{1}{g(A, 0)(e^{\alpha L_0} - 1)} \int_0^{L_0} \frac{p(s)}{W(s)} f(x_0(s), x_0'(s), 0)\, ds.$$

THEOREM 4.1. *In the equation*

(4.1) $$x'' + g(x, x') = \varepsilon f(x, x', \varepsilon)$$

let $g(x, x') \in C^1$ *and* $f(x, x', \varepsilon) \in C^1$. *Let the equation*

(4.2) $$x'' + g(x, x') = 0$$

have the periodic solution $x_0(t)$ *which is nonconstant and of least period* L_0. Let $x_0(0) = A$, $x_0'(0) = 0$. *Let*

(4.13) $$\alpha = \frac{1}{L_0} \int_0^{L_0} g_{x'}(x_0(s), x_0'(s))\, ds \neq 0.$$

Then for sufficiently small ε *there exists a unique periodic solution of* (4.1) *satisfying* $x(0) = A + \xi(\varepsilon)$, $x'(0) = 0$, *of least period* $L_0 + \tau(\varepsilon)$, *where* τ *and* ξ *are* C^1 *functions of* ξ, *and at* $\varepsilon = 0$

$$\frac{d\tau}{d\varepsilon} = \int_0^{L_0} \frac{q(s)}{W(s)} f(x_0(s), x_0'(s), 0)\, ds \equiv \tau'(0)$$

(4.18)

$$\frac{d\xi}{d\varepsilon} = \frac{1}{g(A, 0)(e^{\alpha L_0} - 1)} \int_0^{L_0} \frac{p(s)}{W(s)} f(x_0(s), x_0'(s), 0)\, ds \equiv \xi'(0).$$

The periodic solution has the form

(4.19) $$x = x_0\left(\frac{L_0 t}{L_0 + \tau(\varepsilon)}\right) + \varepsilon x_1\left(\frac{L_0 t}{L_0 + \tau(\varepsilon)}\right) + o(\varepsilon)$$

where $x_1(s)$ *is that unique solution of period* L_0 *of*

(4.8) $$\begin{aligned} y'' + g_{x'}(x_0(s), x_0'(s))y' + g_x(x_0(s), x_0'(s))y \\ = f(x_0(s), x_0'(s), 0) - \frac{2\tau'(0)}{L_0} g(x_0(s), x_0'(s)) \\ + \frac{\tau'(0)}{L_0} g_{x'}(x_0(s), x_0'(s))x_0'(s) \end{aligned}$$

satisfying $y'(0) = 0$, *namely*

$$x_1(s) = \xi'(0)\varphi(s) + \frac{\tau'(0)}{L_0} sp(s) + x_\varepsilon(s). \tag{4.20}$$

Note: In (4.20), $\varphi(s)$ is as defined in (4.15); $x_\varepsilon(s)$ is an abbreviation for $x_\varepsilon(s, 0, 0)$ as given by (4.16).

Proof of Theorem 4.1. The existence has already been proved by the discussion preceding the theorem. The form (4.19) of the solution was also discussed earlier. It is readily verified that $x_1(s)$ as given by (4.20) is a solution of (4.8) which satisfies $x_1(0) = x_1(L_0) = \xi'(0)$, $x_1'(0) = x_1'(L_0) = 0$. Equation (4.8) is derived by making the change of independent variable (4.5) in (4.1), and substituting the expression $x = x_0(s) + \varepsilon x_1(s) + o(\varepsilon)$ into the transformed equation. The equation for $x_1(s)$ is then found to be (4.8).

Remark: The solution (4.19) can be written in powers of ε in the form

$$x = x_0(t) + \varepsilon \tilde{x}_1(t) + o(\varepsilon)$$

where

$$\tilde{x}_1(t) = \xi'(0)\varphi(t) + x_\varepsilon(t).$$

The function $\tilde{x}_1(t)$ is not in general periodic.

The investigation of stability of the periodic solution found in Theorem 4.1 is very simple.

THEOREM 4.2. *The periodic solution found in Theorem* 4.1 *is asymptotically orbitally stable if* $\alpha > 0$, *and is unstable if* $\alpha < 0$.

Proof of Theorem 4.2. The variation equation (4.11) becomes equation (4.12) for $\varepsilon = 0$. Equation (4.11) has one characteristic multiplier equal to +1 for all ε. Now the second characteristic multiplier of (4.12) is $e^{-\alpha L_0}$, whose absolute value is less than or greater than 1 according as α is positive or negative. Hence by continuity, the second characteristic multiplier of (4.11) will be greater or less than 1 in absolute value for small ε according as α is positive or negative. This completes the proof.

Remark: When $\alpha < 0$, the solution is asymptotically orbitally stable as $t \to -\infty$.

Some of the results of this section follow directly from Theorems 2.1 and 2.2, Chapter 14 of [3]. Our theorems give constructive formulas ((4.18) and (4.20)) which show the form of the solutions found.

4.3. Existence and Stability in Cases III and IV. We assume in the present section that both characteristic multipliers of the variation equation (4.12) are equal to +1. This assumption implies that

$$\int_0^{L_0} g_{x'}(x_0(t), x_0'(t))\,dt = 0. \tag{4.21}$$

As a fundamental set of solutions of the variation equation (4.12) we choose $p(t) \equiv x_0'(t)$ and $q(t)$ determined from the initial conditions $q(0) = 1/g(A, 0)$, $q'(0) = 0$. These are the same solutions as used in Sections 3.3 and 3.4. We have

$$q(t) = Ktp(t) + r(t),$$

where $r(t)$ has period L_0. The coefficient K is different from zero in Case III and is equal to zero in Case IV. Note that this makes $q(t)$ periodic in Case IV but not in Case III. The values of $p(t)$, $p'(t)$, $q(t)$, and $q'(t)$ at $t = 0$ and at $t = L_0$ are

$$p(0) = p(L_0) = 0, \qquad p'(0) = p'(L_0) = -g(A, 0),$$

$$q(0) = q(L_0) = \frac{1}{g(A, 0)}, \qquad q'(0) = 0, \quad q'(L_0) = -KLg(A, 0).$$

The Wronskian, $W(t) = p(t)q'(t) - p'(t)q(t)$ satisfies $W(0) = 1$, and is given by

$$W(t) = e^{-\int_0^t g_{x'}(x_0(s), x_0'(s))\,ds}$$

In Cases III and IV, $W(t)$ is periodic with period L_0.

The partial derivatives of $x(t, \xi, \varepsilon)$ for $\xi = \varepsilon = 0$ are given by

$$x_\xi(t, 0, 0) = g(A, 0)q(t),$$

$$x_\varepsilon(t, 0, 0) = \int_0^t \frac{p(s)q(t) - q(s)p(t)}{W(s)} f(x_0(s), x_0'(s), 0)\,ds.$$

With the use of these formulas the first derivatives of $F(\tau, \xi, \varepsilon)$ and $G(\tau, \xi, \varepsilon)$

at $(0, 0, 0)$ are found to be

$$F_\tau(0,0,0) = x'(L_0,0,0) = x_0'(L_0) = 0$$

$$G_\tau(0,0,0) = x''(L_0,0,0) = x_0''(L_0) = -g(A,0)$$

$$F_\xi(0,0,0) = x_\xi(L_0,0,0) - 1 = g(A,0)q(L_0) - 1 = 0$$

$$G_\xi(0,0,0) = x_\xi'(L_0,0,0) = g(A,0)q'(L_0) = -KL_0 g(A,0)^2$$

$$F_\varepsilon(0,0,0) = x_\varepsilon(L_0,0,0) = \frac{1}{g(A,0)} \int_0^{L_0} \frac{p(s)}{W(s)} f(x_0(s), x_0'(s), 0)\,ds,$$

$$G_\varepsilon(0,0,0) = x_\varepsilon'(L_0,0,0) = g(A,0) \int_0^{L_0} \frac{q(s)}{W(s)} f(x_0(s), x_0'(s), 0)\,ds$$

$$- KL_0 g(A,0) \int_0^{L_0} \frac{p(s)}{W(s)} f(x_0(s), x_0'(s), 0)\,ds.$$

It is clear from the above that the Jacobian of F and G with respect to τ and ξ is zero at $(0, 0, 0)$, the matrix being of rank 1. Unless the augmented matrix

$$\left\| \begin{matrix} F_\tau & F_\xi & F_\varepsilon \\ G_\tau & G_\xi & G_\varepsilon \end{matrix} \right\| \tag{4.22}$$

also has rank 1 at $(0, 0, 0)$, there will be no periodic solution of the form sought. But the condition that (4.22) have rank 1 is

$$\int_0^{L_0} \frac{p(s)}{W(s)} f(x_0(s), x_0'(s), 0)\,ds = 0. \tag{4.23}$$

When (4.23) holds, the first partial derivatives of F and G at $(0, 0, 0)$ are given by

$$\begin{aligned} &F_\tau = 0, && F_\xi = 0, && F_\varepsilon = 0, \\ &G_\tau = -g(A,0), && G_\xi = -KL_0 g(A,0)^2, && G_\varepsilon = g(A,0) \int_0^{L_0} \frac{q(s)}{W(s)} f(x_0(s), x_0'(s))\,ds. \end{aligned} \tag{4.24}$$

We conclude from (4.24) that when τ and ξ have been found as functions of ε, the derivatives $d\tau/d\varepsilon$ and $d\xi/d\varepsilon$ at $\varepsilon = 0$ will satisfy

$$\frac{d\tau}{d\varepsilon} + KL_0 g(A, 0)\frac{d\xi}{d\varepsilon} = \int_0^{L_0} \frac{q(s)}{W(s)} f(x_0(s), x_0'(s), 0)\, ds. \tag{4.25}$$

We can illustrate the necessity of condition (4.23) with some examples. We take two different cases with Duffing's equation, where $g(x) = k^2x + \beta x^3$. Every solution of

$$x'' + k^2x + \beta x^3 = 0$$

is periodic, and all are nonconstant except for $x \equiv 0$. (We assume that $\beta > 0$.) The related variation equation is in Case III for $x_0(t) \neq 0$. We consider first the case with $f(x, x', \varepsilon) \equiv x'$. This will correspond to the differential equation

$$x'' - \varepsilon x' + k^2x + \beta x^3 = 0,$$

all solutions of which approach zero as $t \to -\infty$ when ε is positive, and as $t \to +\infty$ when ε is negative. Hence the only possible periodic solution is $x \equiv 0$. We find that this is indicated by the fact that condition (4.23) is not satisfied for any $x_0(t)$ other than $x_0(t) \equiv 0$. Indeed, for this case $W(t) \equiv 1$ since $g(x, x')$ is independent of x', and $p(t) = f(x_0(t), x_0(t), 0) = x_0'(t)$. Thus condition (4.23) becomes

$$\int_0^{L_0} x_0'(t)^2\, dt = 0,$$

which implies that $x_0(t)$ is constant and hence identically zero.

As a second example, suppose that $f(x, x', \varepsilon)$ is independent of x'. The differential equation is then of the form

$$x'' + h(x, \varepsilon) = 0,$$

and all solutions are periodic for sufficiently small ε. In this case we find that condition (4.23) is always satisfied. The condition is:

$$\int_0^{L_0} x_0'(t) f(x_0(t), x_0'(t), 0)\, dt = 0,$$

and this is true since f is independent of x_0' the integral being zero by the periodicity of the indefinite integral of the integrand.

To solve the system $F = G = 0$ for τ and ξ as functions of ε for small ε we solve the equation $G(\tau, \xi, \varepsilon) = 0$ for τ as a function of ξ and ε. This can be done since $G_\tau(0, 0, 0) = -g(A, 0) \neq 0$. If the result is called $\tau = H(\xi, \varepsilon)$, we define $J(\xi, \varepsilon)$ by

$$J(\xi, \varepsilon) = F(H(\xi, \varepsilon), \xi, \varepsilon).$$

We then solve $J(\xi, \varepsilon) = 0$ for ξ as a function of ε for small ε. We can then find τ as a function of ε from

$$\tau = H(\xi(\varepsilon), \varepsilon).$$

The derivative $d\tau/d\varepsilon$ at $\varepsilon = 0$ can be computed from (4.25), using the then known value of $d\xi/d\varepsilon$ at $\varepsilon = 0$.

The first and second partial derivatives of $J(\xi, \varepsilon)$ at $(0, 0)$ can be computed from the first and second partial derivatives of F and G at $(0, 0, 0)$. The computation is as in earlier sections of this paper, and the results are

$$J_\xi = 0, \qquad J_\varepsilon = 0,$$

$$J_{\xi\xi} = KL_0 g(A, 0) g_{x'}(A, 0) - g(A, 0) \int_0^{L_0} [g_{xx'}(x_0(s), x_0'(s)) q(s)$$
$$+ g_{x'x'}(x_0(s), x_0'(s)) q'(s)] ds.$$

$$J_{\xi\varepsilon} = -g_{x'}(A, 0) \int_0^{L_0} \frac{q(s)}{W(s)} f(x_0(s), x_0'(s), 0)\, ds$$
$$- \int_0^{L_0} [g_{xx'}(x_0(s), x_0(s)) x_\varepsilon(s) + g_{x'x'}(x_0(s), x_0'(s)) x_\varepsilon'(s)$$

$$- f_{x'}(x_0(s), x_0'(s), 0)] ds \tag{4.26}$$

$$J_{\varepsilon\varepsilon} = \frac{2}{g(A, 0)} \int_0^{L_0} \frac{p(s)}{W(s)} f_\varepsilon(x_0(s), x_0'(s), 0)\, ds$$
$$- \frac{1}{g(A, 0)} \int_0^{L_0} \frac{x_\varepsilon(s)}{W(s)} g_{x'}(x_0(s), x_0'(s)) f(x_0(s), x_0'(s), 0)\, ds$$
$$- \frac{1}{g(A, 0)} \int_0^{L_0} \frac{p(s) x_\varepsilon'(s) - p'(s) x_\varepsilon(s)}{W(s)} [g_{xx'}(x_0(s), x_0'(s)) x_\varepsilon(s)$$
$$+ g_{x'x'}(x_0(s), x_0'(s)) x_\varepsilon'(s) - 2f_{x'}(x_0(s), x_0'(s), 0)] ds.$$

Remark: In case both g and f are independent of x', we are in the situation of the second example above, where all solutions of the perturbed equation are periodic. The period will not be independent of ε, but since any small ξ will do for small ε, the function $J(\xi, \varepsilon)$ will vanish identically in this case. Note that when $f_{x'}$ and $g_{x'}$ vanish identically, all the terms in the expressions in (4.26) vanish except for the first integral in $J_{\varepsilon\varepsilon}$. This integral vanishes also, since $W(s) \equiv 1$, and the integrand $x_0'(s) f_\varepsilon(x_0(s), x_0'(s), 0)$ is the derivative of a periodic function.

We now have $J(\xi, \varepsilon)$ in the form

$$J(\xi, \varepsilon) = \frac{1}{2} J_{\xi\xi} \xi^2 + J_{\xi\varepsilon} \xi\varepsilon + \frac{1}{2} J_{\varepsilon\varepsilon} \varepsilon^2 + \text{higher order terms.} \tag{4.27}$$

As we have seen earlier, the equation $J(\xi, \varepsilon) = 0$ may be solved for ξ as a function of ε for small ε with $d\xi/d\varepsilon = m$ at $\varepsilon = 0$, if m is a simple real root of the equation

$$\frac{1}{2} J_{\xi\xi} m^2 + J_{\xi\varepsilon} m + \frac{1}{2} J_{\varepsilon\varepsilon} = 0. \tag{4.28}$$

Now (4.28) has one or two simple real roots when the discriminant

$$J_{\xi\varepsilon}^2 - J_{\xi\xi} J_{\varepsilon\varepsilon} > 0,$$

according as $J_{\xi\xi} = 0$ or $J_{\xi\xi} \neq 0$. Assuming the discriminant to be positive, then, we can complete the computation and determine τ as a function of ε. The results are summarized in the following theorem.

THEOREM 4.3. *In the differential equation*

$$x'' + g(x, x') = \varepsilon f(x, x', \varepsilon) \tag{4.1}$$

let $g(x, x') \in C^2$ *and* $f(x, x', \varepsilon) \in C^1$. *Let the unperturbed equation*

$$x'' + g(x, x') = 0 \tag{4.2}$$

have the nonconstant solution $x_0(t)$ *of least period* L_0. *Let both characteristic multipliers of the variation equation*

$$y'' + g_{x'}(x_0(t), x_0'(t)) y' + g_x(x_0(t), x_0'(t)) y = 0 \tag{4.12}$$

be equal to $+1$. *Let*

$$(4.23)\qquad \int_0^{L_0} \frac{p(s)}{W(s)} f(x_0(s), x_0'(s), 0)\, ds = 0.$$

Let the discriminant of (4.28) *be positive.*

Then, according as $J_{\xi\xi} = 0$ *or* $J_{\xi\xi} \neq 0$, *there exist one or two periodic solutions of* (4.1) *which reduce to* $x_0(t)$ *for* $\varepsilon = 0$. *These periodic solutions satisfy the initial conditions* $x(0) = A + m\varepsilon + o(\varepsilon)$, $x'(0) = 0$, *where* m *is a simple real root of* (4.28), *and have period* $L_0 + \tau(\varepsilon)$, *where* $\tau(0) = 0$ *and* $\tau'(0)$ *is given by*

$$(4.29)\qquad \tau'(0) = -KL_0 g(A, 0)\, m + \int_0^{L_0} \frac{q(s)}{W(s)} f(x_0(s), x_0'(s), 0)\, ds.$$

The periodic solutions have the form

$$(4.30)\qquad x = x_0\left(\frac{L_0 t}{L_0 + \tau(\varepsilon)}\right) + \varepsilon x_1\left(\frac{L_0 t}{L_0 + \tau(\varepsilon)}\right) + o(\varepsilon)$$

where $x_1(s)$ *is that periodic solution of* (4.8) *which satisfies* $x_1'(0) = 0$, *namely*

$$(4.31)\qquad x_1(s) = mg(A, 0)\, q(t) + x_\varepsilon(s) + \frac{\tau'(0)}{L_0}\, sp(s).$$

Proof of Theorem 4.3. The existence of a periodic solution corresponding to each simple root, m, of equation (4.28) has been proved in the discussion preceding the statement of the theorem. From its very construction, $m = d\xi/d\varepsilon$ at $\varepsilon = 0$, which establishes the formula for $x(0) + A + \xi$. The formula for $\tau'(0)$ follows from (4.25).

When the change of variable (4.5) is made, there results the differential equation (4.6), which has solutions of period L_0 for small ε, of the form,

$$x = x_0(s) + \varepsilon x_1(s) + o(\varepsilon),$$

where $x_1(s)$ is periodic of period L_0, and is a solution of (4.8). It is readily verified that (4.31) is the solution in question. This completes the proof.

Knowing the existence of the periodic solutions or solution found in Theorem 4.3, we can use a formal procedure to determine $\tau'(0)$ and $\xi'(0)$. In the original differential equation

(4.1) $$x'' + g(x, x') = \varepsilon f(x, x', \varepsilon)$$

make the double substitution

(4.32) $$\begin{aligned} x &= x_0 + \varepsilon x_1 + \varepsilon^2 x_2 + \dots \\ t &= s(1 + \gamma_1 \varepsilon + \gamma_2 \varepsilon^2 + \dots) \end{aligned}$$

After the substitution (4.32) is made, coefficients of like powers of ε are equated to give a sequence of equations which determine in turn $x_0, x_1, x_2, \dots$ as functions of s. The equation satisfied by x_0 is, of course, (4.2), and we use the solution $x_0(s)$ that we have been using. The orthogonality condition for existence of a periodic $x_1(s)$ is found to be just (4.23). The periodic solution which satisfies also $x'(0) = 0$ is

(4.33) $$x_1(s) = x_\varepsilon(s) + \gamma_1 s p(s) + \frac{q(s)}{KL_0} \int_0^{L_0} \frac{q(u)}{W(u)} f(x_0(u), x_0'(u), 0)\, du - \frac{\gamma_1 q(s)}{K}$$

Now x has period L_0 in s, so that it has period

$$L_0 + \tau(\varepsilon) = L_0 + L_0\gamma_1 \varepsilon + L_0 \gamma_2 \varepsilon^2 + \dots$$

in t. Thus $\tau'(0) = L_0 \gamma_1$. If we denote the initial value of x_1, $x_1(0) = \xi'(0)$, by m, we find from (4.33)

(4.34) $$m = \frac{1}{KL_0 g(A, 0)} \int_0^{L_0} \frac{q(u)}{W(u)} f(x_0(u), x_0'(u), 0)\, du - \frac{\gamma_1}{Kg(A, 0)}$$

and since $\gamma_1 L_0 = \tau'(0)$, (4.34) is the same as (4.29).

Now rewrite equation (4.33) as

(4.35) $$x_1(s) = x_\varepsilon(s) + \gamma_1 s p(s) + m g(A, 0) q(s),$$

and use this in the differential equation for $x_2(s)$. The orthogonality condition for the existence of a periodic $x_2(s)$ is the vanishing of a function of m and γ_1. If γ_1 is eliminated by the use of (4.34), the orthogonality condition will reduce to an equation for m which is equivalent to (4.28). Thus in the case considered in Theorem 4.3, the quantities $m = \xi'(0)$ and $\tau'(0) = L_0 \gamma_1$, are determined ultimately by the orthogonality condition for the existence of a periodic $x_2(s)$.

We now consider the case in which the discriminant of (4.28) is not positive.

If the discriminant is negative, equation (4.28) has no real solutions, and no real periodic solution of the type sought exists. If the discriminant is zero, with $J_{\xi\xi} = J_{\xi\varepsilon} = 0$, $J_{\varepsilon\varepsilon} \neq 0$, equation (4.28) has no solutions at all, and no solution of the type sought exists. However under certain circumstances we can find a solution of the type we seek when the discriminant vanishes but $J_{\xi\xi} \neq 0$, and in the case that the left member of (4.28) vanishes identically.

The techniques of study were used in Theorem 3.15 in Section 3.4. The third derivatives of $J(\xi, \varepsilon)$ will be required for this discussion. These can be computed from the third derivatives of F and G as earlier. Since the expressions are long and cumbersome, we omit them. We have then

$$J(\xi, \varepsilon) = \frac{1}{2}J_{\xi\xi}\xi^2 + J_{\xi\varepsilon}\xi\varepsilon + \frac{1}{2}J_{\varepsilon\varepsilon}\varepsilon^2$$

$$+ \frac{1}{6}J_{\xi\xi\xi}\xi^3 + \frac{1}{2}J_{\xi\xi\varepsilon}\xi^2\varepsilon + \frac{1}{2}J_{\xi\varepsilon\varepsilon}\xi\varepsilon^2 + \frac{1}{6}J_{\varepsilon\varepsilon\varepsilon}\varepsilon^3$$

+ higher order terms.

Consider first the case in which $J_{\xi\xi} \neq 0$, but

$$J_{\xi\varepsilon}^2 - J_{\xi\xi}J_{\varepsilon\varepsilon} = 0.$$

In this case the second degree terms in $J(\xi, \varepsilon)$ become simply

$$\frac{1}{2}J_{\xi\xi}(\xi - m\varepsilon)^2$$

where m is the double root of equation (4.26). Let $\xi = (v + m)\varepsilon$. We then find

$$J(\xi, \varepsilon) = \frac{1}{2}\varepsilon^2 J_{\xi\xi}v^2 + \varepsilon^3(J_3 + \text{terms involving } v) \tag{4.36}$$

$$+ \text{ higher order terms},$$

where

$$J_3 = \frac{1}{6}J_{\xi\xi\xi}m^3 + \frac{1}{2}J_{\xi\xi\varepsilon}m^2 + \frac{1}{2}J_{\xi\varepsilon\varepsilon}m + \frac{1}{6}J_{\varepsilon\varepsilon\varepsilon}.$$

Now since we are not interested in $\varepsilon = 0$, we divide (4.36) through by ε^2, obtaining as the equation to be solved

$$\frac{1}{2}J_{\xi\xi}v^2 + \varepsilon(J_3 + \text{terms involving } v) + \text{higher order terms} = 0. \tag{4.37}$$

If $J_3 \neq 0$, we can solve (4.37) for v as a function of ε, either when ε is positive, or when ε is negative, but not both. If $J_3 = 0$, higher derivatives of F and G at $(0, 0, 0)$ are required, and we do not consider the situation. If ε is positive, and J_3 and $J_{\xi\xi}$ have opposite signs, then

$$v = \pm \sqrt{\frac{-2J_3}{J_{\xi\xi}}}\ \varepsilon^{\frac{1}{2}} + \ldots,$$

while if ε is negative, and J_3 and $J_{\xi\xi}$ have the same sign,

$$v = \pm \sqrt{\frac{2J_3}{J_{\xi\xi}}}\ (-\varepsilon)^{\frac{1}{2}} + \ldots .$$

Returning to the variable ξ, we find

$$\xi = m\varepsilon \pm \sqrt{\frac{-2J_3}{J_{\xi\xi}}}\ \varepsilon^{\frac{3}{2}} + o(\varepsilon^{\frac{3}{2}}) \tag{4.38}$$

when $J_{\xi\xi}$ and J_3 have opposite signs, the solutions being valid for positive ε and not for negative ε, while

$$\xi = m\varepsilon \pm \sqrt{\frac{2J_3}{J_{\xi\xi}}}\ (-\varepsilon)^{\frac{3}{2}} + o(|\varepsilon|^{\frac{3}{2}}) \tag{4.39}$$

when $J_{\xi\xi}$ and J_3 have the same sign, the solutions being valid for negative ε, and not for positive ε.

We have thus solved $J(\xi, \varepsilon) = 0$ for ξ as a function of ε, there being either two solutions for positive ε and none for negative ε or else two solutions for negative ε and none for positive ε. In each case $\xi'(0) = m$, the double root of (4.28).

We now consider the case in which all three of $J_{\xi\xi}$, $J_{\xi\varepsilon}$ and $J_{\varepsilon\varepsilon}$ are zero. In this case we find

$$J(\xi, \varepsilon) = \frac{1}{6} J_{\xi\xi\xi}\xi^3 + \frac{1}{2} J_{\xi\xi\varepsilon}\xi^2\varepsilon + \frac{1}{2} J_{\xi\varepsilon\varepsilon}\xi\varepsilon^2 + \frac{1}{6} J_{\varepsilon\varepsilon\varepsilon}\varepsilon^3$$

$$+ \text{ higher order terms.}$$

If m is a simple real root of

$$\frac{1}{6}J_{\xi\xi\xi}m^3 + \frac{1}{2}J_{\xi\xi\varepsilon}m^2 + \frac{1}{2}J_{\xi\varepsilon\varepsilon}m + \frac{1}{6}J_{\varepsilon\varepsilon\varepsilon} = 0, \tag{4.40}$$

we have a solution of $J(\xi, \varepsilon) = 0$ of the form

$$\xi = m\varepsilon + o(\varepsilon). \tag{4.41}$$

Multiple roots of (4.40) require knowledge of higher derivatives of F and G, and we do not consider this case here.

We summarize the existence results in the following theorem.

THEOREM 4.4. *Let all the hypotheses of Theorem 4.3 through (4.23) be satisfied. In addition let* $g(x, x') \in C^3$ *and* $f(x, x', \varepsilon) \in C^2$. *Finally let one of the two following situations occur:*

(a) *Equation (4.28) has a double root,* m, *with the quantity* $J_3 \neq 0$.

(b) *All three second derivatives of* $J(\xi, \varepsilon)$ *are zero at* $\xi = \varepsilon = 0$, *and equation (4.40) has a simple root* m.

Then in case (a) there exist two periodic solutions of equation (4.2) for small ε, *both existing for positive* ε *and not for negative* ε *if* $J_{\xi\xi}$ *and* J_3 *have opposite signs, and both existing for negative* ε *and not for positive* ε *if* $J_{\xi\xi}$ *and* J_3 *have the same sign. In case (b) there exists one periodic solution of equation (4.2) for small* ε *for each simple real root of equation (4.40). These periodic solutions have all the properties listed in Theorem 4.3.*

Proof of Theorem 4.4. The proof of existence has been given in the discussion preceding the statement of the theorem. The proof that the properties listed in Theorem 4.3 hold is identical with the corresponding part of the proof of that theorem.

We now turn to the question of stability of the periodic solutions found in Theorems 4.3 and 4.4. As has been remarked earlier, we must investigate the second characteristic multiplier of the variational equation in the vicinity of $\varepsilon = 0$. If this characteristic multiplier is less than 1, we have orbital stability. If it is greater than 1, we have instability.

Let $x = x(t, \varepsilon)$ denote a periodic solution found in Theorem 4.3 or Theorem 4.4. Let its period be $L = L(\varepsilon) = L_0 + \tau(\varepsilon)$. The variation equation that we must study is

(4.42) $$y'' + g_{x'}(x, x')y' + g_x(x, x')y = \varepsilon f_x(x, x', \varepsilon)y + \varepsilon f_{x'}(x, x', \varepsilon)y'.$$

Let $\varphi(t, \varepsilon)$ and $\psi(t, \varepsilon)$ be those solutions of (4.42) which satisfy $\varphi(0, \varepsilon) = \psi'(0, \varepsilon) = 1$, $\varphi'(0, \varepsilon) = \psi(0, \varepsilon) = 0$. When $\varepsilon = 0$, equation (4.42) reduces to (4.12), and the solutions φ and ψ become $\varphi(t, 0) = g(A, 0)q(t)$, $\psi(t, 0) = -p(t)/g(A, 0)$.

Since $x'(t, \varepsilon)$ is a solution of (4.42), and since $x'(0, \varepsilon) = 0$, it follows that $\psi(t, \varepsilon)$ is a constant multiple of $x'(t, \varepsilon)$ and is therefore periodic of period L.

The characteristic multipliers of (4.42) are the eigenvalues of the matrix

(4.43) $$\begin{Vmatrix} \varphi(L, \varepsilon) & \psi(L, \varepsilon) \\ \varphi'(L, \varepsilon) & \psi'(L, \varepsilon) \end{Vmatrix}$$

Since $\psi(t, \varepsilon)$ has period L, we have $\psi(L, \varepsilon) = 0$, $\psi'(L, \varepsilon) = 1$. Therefore the two characteristic multipliers are 1 and $\varphi(L, \varepsilon)$, and the critical quantity for stability is $\varphi(L, \varepsilon)$.

Recall the solution $x(t, \xi, \varepsilon)$ of equation (4.1). The periodic solution $x(t, \varepsilon)$ is exactly $x(t, \xi(\varepsilon), \varepsilon)$, where $\xi(\varepsilon)$ is the value of ξ found by solving $J(\xi, \varepsilon) = 0$. The partial derivative $x_\xi(t, \xi, \varepsilon)$ evaluated at the periodic solution $x(t, \varepsilon)$ is that solution of (4.42) with $y(0) = 1$, and $y'(0) = 1$. But this is exactly $\varphi(t, \varepsilon)$, so that

$$\varphi(L, \varepsilon) = \frac{\partial}{\partial \xi} x(t, \xi, \varepsilon) \Big|_{t=L,\ \xi = \xi(\varepsilon)}.$$

Now

$$x(L_0 + \tau, \xi, \varepsilon) = F(\tau, \xi, \varepsilon) + A + \xi$$

so that

$$\frac{\partial}{\partial \xi} x(L_0 + \tau, \xi, \varepsilon) = F_\xi(\tau, \xi, \varepsilon) + 1,$$

and

$$(L, \varepsilon) = \frac{\partial}{\partial \xi} x(L_0 + \tau, \xi, \varepsilon) \Big|_{\tau = \tau(\varepsilon),\ \xi = \xi(\varepsilon)}$$

$$= F_\xi(\tau(\varepsilon), \xi(\varepsilon), \varepsilon) + 1.$$

We now introduce the function $J(\xi, \varepsilon)$, and establish that

$$F_\xi(\tau(\varepsilon), \xi(\varepsilon), \varepsilon) = J_\xi(\xi(\varepsilon), \varepsilon).$$

To see this, recall that

$$J(\xi, \varepsilon) = F(H(\xi, \varepsilon), \xi, \varepsilon)$$

where $\tau = H(\xi, \varepsilon)$ is the result of solving the equation $G(\tau, \xi, \varepsilon) = 0$ for τ as a function of ξ and ε. Thus

$$J_\xi(\xi, \varepsilon) = F_\tau(H(\xi, \varepsilon), \xi, \varepsilon) H_\xi(\xi, \varepsilon) + F_\xi(H(\xi, \varepsilon), \xi, \varepsilon).$$

Now when $\xi = \xi(\varepsilon)$, we find $\tau(\varepsilon)$ from

$$\tau(\varepsilon) = H(\xi(\varepsilon), \varepsilon).$$

Therefore

$$J_\xi(\xi, (\varepsilon), \varepsilon) = F_\tau(\tau(\varepsilon), \xi(\varepsilon), \varepsilon) H_\xi(\xi(\varepsilon), \varepsilon) + F_\xi(\tau(\varepsilon), \xi(\varepsilon), \varepsilon).$$

But

$$F_\tau(\tau, \xi, \varepsilon) = x'(L_0 + \tau, \xi, \varepsilon),$$

so that

$$F_\tau(\tau(\varepsilon), \xi(\varepsilon), \varepsilon) = x'(L, \xi(\varepsilon), \varepsilon).$$

Now $x(\tau, \xi(\varepsilon), \varepsilon) \equiv x(\tau, \varepsilon)$, so that

$$x'(L, \xi(\varepsilon), \varepsilon) = x'(L, \varepsilon) = x'(0, \varepsilon) = 0.$$

Therefore

$$J_\xi(\xi(\varepsilon), \varepsilon) = F_\xi(\tau(\varepsilon), \xi(\varepsilon), \varepsilon).$$

In summary we have found that the second characteristic multiplier, which determines stability, is given by

$$\varphi(L, \varepsilon) = 1 + J_\xi(\xi(\varepsilon), \varepsilon).$$

We now examine the various cases which occur in Theorems (4.3) and (4.4). In Theorem (4.3), we found

$$\xi(\varepsilon) = m\varepsilon + o(\varepsilon),$$

where m is a simple real root of equation (4.28). Also

$$J_\xi(\xi, \varepsilon) = J_{\xi\xi}\xi + J_{\xi\varepsilon}\varepsilon + \text{higher order terms},$$

$$J_\xi(\xi(\varepsilon), \varepsilon) = \varepsilon(J_{\xi\xi}m + J_{\xi\varepsilon}) + o(\varepsilon).$$

Now since m is a simple root of (4.28), $J_{\xi\xi}m + J_{\xi\varepsilon} \neq 0$. Thus the second characteristic multiplier is

$$\varphi(L, \varepsilon) = 1 + \varepsilon(J_{\xi\xi}m + J_{\xi\varepsilon}) + o(\varepsilon),$$

and we have for small ε orbital stability or instability according as $\varepsilon(J_{\xi\xi}m + J_{\xi\varepsilon})$ is negative or positive. Note that when $J_{\xi\xi} \neq 0$, and there are two possible values of m, that one of the solutions will be orbitally stable, while the second is unstable.

There were two cases in Theorem 4.4. We found when $J_{\xi\xi}$ and J_3 are of opposite sign that we have two solutions $\xi(\varepsilon)$ for positive ε and none for negative ε, with

$$\xi(\varepsilon) = m\varepsilon \pm \sqrt{\frac{-2J_3}{J_{\xi\xi}}}\,\varepsilon^{\frac{3}{2}} + o(\varepsilon^{\frac{3}{2}})$$

where m is the double root of equation (4.28). In this case

$$J(\xi, \varepsilon) = \frac{1}{2} J_{\xi\xi}(\xi - m\varepsilon)^2 + \varepsilon^3 (J_3 + 0(\frac{\xi}{\varepsilon} - m))$$

$$+ \text{ higher order terms}$$

$$J_\xi(\xi, \varepsilon) = J_{\xi\xi}(\xi - m\varepsilon) + \varepsilon^2 0(\frac{\xi}{\varepsilon} - m) + \text{ higher order terms}$$

$$J_\xi(\xi(\varepsilon), \varepsilon) = \pm \operatorname{sgn} J_{\xi\xi} \sqrt{-2J_3 J_{\xi\xi}}\, \varepsilon^{\frac{3}{2}} + o(\varepsilon^{\frac{3}{2}}).$$

Thus the second characteristic multiplier is

$$\varphi(L, \varepsilon) = 1 \pm \operatorname{sgn} J_{\xi\xi} \sqrt{-2J_3 J_{\xi\xi}}\, \varepsilon^{\frac{3}{2}} + o(\varepsilon^{\frac{3}{2}}).$$

Hence of the two periodic solutions one is orbitally stable, and the other is unstable.

In a similar manner, when J_3 and $J_{\xi\xi}$ have the same sign, we have for small negative ε that the second characteristic multiplier is given by

$$\varphi(L, \varepsilon) = 1 \pm \operatorname{sgn} J_{\xi\xi} \sqrt{2J_3 J_{\xi\xi}}\, (-\varepsilon)^{\frac{3}{2}} + o(|\varepsilon|^{\frac{3}{2}}),$$

so that again one solution is orbitally stable and the other is unstable.

In the second case of Theorem (4. 4), all second degree terms in $J(\xi, \varepsilon)$ were zero. We still found

$$\xi(\varepsilon) = m\varepsilon + o(\varepsilon),$$

where m is a simple real root of equation (4. 40). By a calculation similar to those above we find that for such a periodic solution, the second characteristic multiplier is given by

$$\varphi(L, \varepsilon) = 1 + \varepsilon^2 (\frac{1}{2} J_{\xi\xi\xi} m^2 + J_{\xi\xi\varepsilon} m + \frac{1}{2} J_{\xi\varepsilon\varepsilon}) + o(\varepsilon^2)$$

so that the corresponding solution is orbitally stable or unstable according as

$$\frac{1}{2} J_{\xi\xi\xi} m^2 + J_{\xi\xi\varepsilon} m + \frac{1}{2} J_{\xi\varepsilon\varepsilon}$$

is negative or positive. (It is not zero, since m is a simple root of (4. 40).)

THEOREM 4. 5. *Let all of the hypotheses of Theorem* 4. 3 *or Theorem* 4. 4 *as the case may be, hold. The periodic solutions found in those theorems have the following stability properties:*

(a) *The periodic solutions found in Theorem* 4.3 *are orbitally stable or unstable according as*

$$\varepsilon(J_{\xi\xi}m + J_{\xi\varepsilon})$$

is negative or positive.

(b) *The periodic solutions found in the first case of Theorem* (4.4) *occur in pairs, one orbitally stable and the other unstable.*

(c) *The periodic solutions found in the second case of Theorem* (4.4) *are orbitally stable or unstable according as*

$$\frac{1}{2}J_{\xi\xi\xi}m^2 + J_{\xi\xi\varepsilon}m + \frac{1}{2}J_{\xi\varepsilon\varepsilon}$$

is negative or positive.

Proof: The proof of the theorem is contained in the discussion which precedes it.

There is another way to see the way that stability is determined by the algebraic sign of $J_\xi(\xi(\varepsilon), \varepsilon)$. If the solution $x(t, \xi, \varepsilon)$ is followed for a time $L_0 + H(\xi, \varepsilon)$, its derivative at that time is again zero. At this time, its position is not $A + \xi$, but $A + \xi + F(H(\xi, \varepsilon), \xi, \varepsilon) = A + \xi + J(\xi, \varepsilon)$. Thus $J(\xi, \varepsilon)$ is the change in displacement of the solution $x(t, \xi, \varepsilon)$ in one circuit of its orbit, from the point $x = A + \xi$, $x' = 0$ around to a neighboring point with $x' = 0$ again. $x(t, \xi, \varepsilon)$ is periodic if this change in displacement, $J(\xi, \varepsilon)$ is zero. If as ξ increases through such a critical value, $\xi(\varepsilon)$, J changes from positive to negative, it is clear that the corresponding periodic solution is stable, since if ξ is too large, $J(\xi, \varepsilon)$ is negative and the displacement is decreased, while if ξ is too small, $J(\xi, \varepsilon)$ is positive, and the displacement is increased. This corresponds to $J_\xi(\xi(\varepsilon), \varepsilon)$ being nonpositive, and stability certainly occurs if $J_\xi(\xi(\varepsilon), \varepsilon)$ is negative. By similar reasoning we have instability of $J_\xi(\xi(\varepsilon), \varepsilon)$ is positive. This is in agreement with the results in Theorem 4.5.

4.4. The Case that the Unperturbed Equation has a Family of Periodic Solutions

Let us now consider the case where the unperturbed equation

$$x'' + g(x, x') = 0 \tag{4.2}$$

has a family of periodic solutions. By this we mean that for some interval

$A_1 < A < A_2$, the solutions of (4.2) with $x(0) = A$, $x'(0) = 0$ are periodic with least period $L(A)$, say. We assume that $g(A, 0) \neq 0$ in this interval, so that none of the periodic solutions is identically constant. The period $L(A)$ will usually vary with A, and will be independent of A only in exceptional cases.

Let us denote by $x_0(t, A)$ the periodic solution of the above family with $x_0(0, A) = A$. The related variational equation will be

$$(4.44) \qquad y'' + g_{x'}(x_0(t, A), x_0'(t, A))y' + g_x(x_0(t, A, x_0'(t, A))y = 0.$$

As before define the fundamental set of solutions $p(t, A)$ and $q(t, A)$ by the conditions $p(t, A) \equiv x_0'(t, A)$, so that $p(0, A) = 0$, $p'(0, A) = -g(A, 0)$, $q(0, A) = 1/g(A, 0)$, $q'(0, A) = 0$. We denote the Wronskian of $p(t, A)$ and $q(t, A)$ by $W(t, A)$. For each value of A, equation (4.44) is in Case III or Case IV, i.e.,

$$(4.45) \qquad \int_0^{L(A)} g_{x'}(x_0(t, A), x_0'(t, A)\,dt = 0.$$

This is because the periodic solutions $x_0(t, A)$ are not isolated. It should perhaps be remarked that although the existence of a family of periodic solutions of (4.2) implies that the variation equation is in Case III or Case IV, the converse implication does not hold. The variation equation (4.12) may be in Case III or Case IV for isolated values of A, with no implication that $x_0(t)$ is embedded in a family of periodic solutions of (4.2). When $L(A)$ is independent of A, equation (4.44) is necessarily in Case IV, but the converse implication does not hold. Equation (4.44) can be in Case IV for isolated values of A without the implication that $L(A)$ is constant.

It will usually be the case that there will be no periodic solution of the perturbed equation

$$(4.1) \qquad x'' + g(x, x') = \varepsilon f(x, x', \varepsilon)$$

which reduces as $\varepsilon \to 0$ to the periodic solution $x_0(t, A)$ of (4.2). We propose to show how the earlier results of this part of the paper can be used to determine those values of A for which (4.1) has periodic solutions reducing to $x_0(t, A)$.

Let $x(t, A, \xi, \varepsilon)$ be that solution of (4.1) which for $t = 0$ has $x = A + \xi$,

$x' = 0$. Note that $x(t, A, 0, 0) \equiv x_0(t, A)$. Define $\tilde{F}(\tau, A, \xi, \varepsilon)$ and $\tilde{G}(\tau, A, \xi, \varepsilon)$ by

$$\tilde{F}(\tau, A, \xi, \varepsilon) = x(L(A) + \tau, A, \xi, \varepsilon) - A - \xi,$$

$$\tilde{G}(\tau, A, \xi, \varepsilon) = x'(L(A) + \tau, A, \xi, \varepsilon),$$

The solution $x(t, A, \xi, \varepsilon)$ will be periodic with period $L(A) + \tau$ if and only if $\tilde{F} = \tilde{G} = 0$.

We shall solve the system $\tilde{F} = \tilde{G} = 0$ for ξ and τ as functions of ε and A. This means we shall find for which A this is possible, and then establish that there are solutions of the form

$$\tau = \tau_1 \varepsilon + o(\varepsilon) \qquad \xi = m\varepsilon + o(\varepsilon),$$

where we can evaluate $d\tau/d\varepsilon$ and $d\xi/d\varepsilon$ at $\varepsilon = 0$. We note that for $\tau = \xi = \varepsilon = 0$, $\tilde{F} = \tilde{G} = 0$.

We use implicit function methods as earlier. One remark that should be made is that because A and ξ enter in the combination $A + \xi$, differentiation of $\tilde{F}$ and $\tilde{G}$ with respect to A is the same as differentiation with respect to ξ.

For fixed A, $\tilde{F}$ and $\tilde{G}$ become the F and G of the earlier sections. Therefore $\tilde{G}_\tau(0, A, 0, 0) = -g(A, 0) \neq 0$, for $A_1 < A < A_2$. We may therefore solve $\tilde{G}(\tau, A, \xi, \varepsilon) = 0$ for τ as a function of A, ξ, and ε. Write the result as $\tau = \tilde{H}(A, \xi, \varepsilon)$, and define $\tilde{J}(A, \xi, \varepsilon)$ as

$$\tilde{J}(A, \xi, \varepsilon) = \tilde{F}(\tilde{H}(A, \xi, \varepsilon), A, \xi, \varepsilon).$$

We are to solve $\tilde{J}(A, \xi, \varepsilon) = 0$ for ξ as a function of A and ε.

Since for $\varepsilon = 0$ there is a family of periodic solutions for $A_1 < A < A_2$, we have

$$(4.46) \qquad \tilde{J}(A, \xi, 0) \equiv 0. \qquad A_1 < A + \xi < A_2$$

This implies in particular that

$$(4.47) \qquad \tilde{J}_\xi(A, \xi, 0) = \tilde{J}_{\xi\xi}(A, \xi, 0) = \tilde{J}_{\xi\xi\xi}(A, \xi, 0) \equiv 0.$$

Therefore in order that we can solve $\tilde{J}(A, \xi, \varepsilon) = 0$ for ξ as a function of ε, we must have $\tilde{J}_\varepsilon(A, 0, 0) = 0$, as a necessary condition. Referring back to Section 4.3, we see that $\tilde{J}_\varepsilon(A, 0, 0)$ is equal to

$$(4.48)\qquad F_1(A) = \frac{1}{g(A, 0)} \int_0^{L(A)} \frac{p(s, A)}{W(s, A)} f(x_0(s, A), x_0'(s, A), 0)\, ds.$$

The necessary condition is then $F_1(A) = 0$, analogous to condition (4.23). Now by Theorem 4.3, a sufficient condition for existence of ξ as a function of ε, along with $\tilde{J}_\varepsilon(A, 0, 0) = 0$ is that $\tilde{J}_{\xi\varepsilon}(A, 0, 0) \neq 0$. But since differentiation with respect to ξ or A is the same, this condition is $F_1'(A) = 0$.

When $F_1'(A) \neq 0$, the stability is determined according to Theorem 4.5 by the algebraic sign of $\varepsilon J_{\xi\varepsilon}(A, 0, 0) = \varepsilon F_1'(A)$. Therefore we can restate Theorem 4.3 and the corresponding part of Theorem 4.5 as follows:

THEOREM 4.6. *In the differential equation*

$$(4.1)\qquad x'' + g(x, x') = \varepsilon f(x, x', \varepsilon)$$

let $g(x, x') \in C^2$ *and* $f(x, x', \varepsilon) \in C^1$. For $\varepsilon = 0$ *let there be a family of periodic solutions* $x_0(t, A)$ $(A_1 < A < A_2)$ *with* $g(A, 0) \neq 0$ *on this range. Let* $F_1(A)$ *be defined by* (4.46).

Then for any A on $A_1 < A < A_2$ *for which* $F_1(A) = 0$ *and* $F_1'(A) \neq 0$, *there exists for sufficiently small* ε *a periodic solution of* (4.1) *of the type studied in this paper, which reduces to* $x_0(t, A)$ *as* $\varepsilon \to 0$. *This solution is orbitallv stable or unstable according as* $\varepsilon F_1'(A)$ *is negative or positive.*

We now consider the case where $F_1(A)$ has a multiple zero. This may possibly involve the identical vanishing of $F_1(A)$. We are able to obtain the result from Theorem 4.4. Of the two cases in Theorem (4.4), only the second can occur here. By (4.47), $\tilde{J}_{\xi\xi}(A, 0, 0) = 0$, so that only case (b) will occur. Of the other two second partial derivatives of $\tilde{J}(A, \xi, \varepsilon)$ at $\xi = \varepsilon = 0$, $\tilde{J}_{\xi\varepsilon}(A, 0, 0) = 0$, because this is $F_1'(A)$, which is zero by hypothesis in the present discussion. Thus the only necessary condition that must be imposed in the present discussion is that

$$F_2(A) \equiv \tilde{J}_{\varepsilon\varepsilon}(A, 0, 0) = 0.$$

According to Theorem 4.4, it is further required that the equation

$$(4.49)\qquad \frac{1}{6}\tilde{J}_{\xi\xi\xi} m^3 + \frac{1}{2}\tilde{J}_{\xi\xi\varepsilon} m^2 + \frac{1}{2}\tilde{J}_{\xi\varepsilon\varepsilon} m + \frac{1}{6}\tilde{J}_{\varepsilon\varepsilon\varepsilon} = 0$$

have a simple real root. In (4.49) the partial derivatives of $\tilde{J}$ in the coefficients are evaluated at $(A, 0, 0)$. By (4.47) the coefficient of m^3 is zero. The coefficient of m^2 is just $\frac{1}{2}F_1''(A)$. Thus (4.49) is either linear or quadratic, depending on whether the zero of $F_1(A)$ is of order higher than two, or is of order precisely two. The coefficient of m in (4.49) is just $\frac{1}{2}F_2'(A)$. The discriminant, which must be positive is

$$\frac{1}{4}\tilde{J}^2_{\xi\varepsilon\varepsilon} - \frac{1}{3}\tilde{J}_{\xi\xi\varepsilon}\tilde{J}_{\varepsilon\varepsilon\varepsilon}. \tag{4.50}$$

The stability can also be found from Theorem (4.5). If m is a simple real root of (4.49), the corresponding periodic solution for small ε is orbitally stable or unstable according as

$$\tilde{J}_{\xi\xi\varepsilon}m + \frac{1}{2}\tilde{J}_{\xi\varepsilon\varepsilon} \tag{4.51}$$

is negative or positive.

THEOREM 4.7. *Let the hypotheses in the first paragraph of Theorem* 4.6 *hold except that* $g(x, x') \in C^3$, $f(x, x', \varepsilon) \in C^2$. *Let* A *be a multiple zero of* $F_1(A)$ *on the range* $A_1 < A < A_2$. *For this value of* A *let*

$$F_2(A) \equiv \tilde{J}_{\varepsilon\varepsilon}(A, 0, 0) = 0,$$

and let the discriminant (4.50) *of equation* (4.49) *be positive.*

Then for sufficiently small ε *there exist one or two periodic solutions of* (4.1) *of the type studied in this paper which reduce to* $x_0(t, A)$ *as* $\varepsilon \to 0$, *according as* $F_1''(A) = 0$ *or* $F_1''(A) \neq 0$. *Any such solution is orbitally stable or unstable according as*

$$\tilde{J}_{\xi\xi\varepsilon}m + \frac{1}{2}\tilde{J}_{\xi\varepsilon\varepsilon} \tag{4.51}$$

is negative or positive, where m *is the corresponding simple real root of* (4.49).

Remark: When there are two periodic solutions, they are of opposite stability. When there is but one periodic solution (as is the case when $F_1(A) \equiv 0$), the stability is determined by the algebraic sign of $F_2'(A)$.

As an example for Theorem 4.6, consider van der Pol's equation

$$(4.52) \qquad x'' + x = \varepsilon(1 - x^2)x'.$$

Here $x_0(t, A) = A \cos t$, and this is valid for all nonzero A. Thus as the range $A_1 < A < A_2$ we may take any open interval of positive values of A. The function $F_1(A)$ is readily computed and found to be

$$F_1(A) = \frac{1}{4}\pi A(4 - A^2).$$

$F_1(A)$ has a simple positive zero at $A = 2$. $\varepsilon F_1'(2) = -2\pi\varepsilon$. Thus by Theorem 4.6 there exists a periodic solution of (4.52) reducing for $\varepsilon = 0$ to $x_0(t, 2) = 2 \cos t$. Moreover this solution is orbitally stable or unstable according as ε is positive or negative. This is in agreement with well known results.

As an example for Theorem 4.7 consider

$$x'' + x = \varepsilon(x^2 - 2xx' + 4xx'^3).$$

For this equation we again have $x_0(t, A) = A \cos t$ and we use for the interval (A_1, A_2) any open interval of positive values. For this equation it is found that $F_1(A) \equiv 0$, and that $F_2(A) = \pi A^3(A^2 - 1)$. Thus there is a periodic solution near to $x_0(t, 1) = \cos t$.

BIBLIOGRAPHY

1. Bogolyubov, N. N., and Metropolskii, Yu. A., *Asymptotic Methods in the Theory of Nonlinear Oscillations*, second edition, Moscow, 1958 (Russian).

2. Cesari, L., *Asymptotic Behavior and Stability Problems in Ordinary Differential Equations*, Springer, Berlin, 1959.

3. Coddington, E.A., and Levinson, N., *Theory of Ordinary Differential Equations*, McGraw-Hill, New York, 1955.

4. Friedrichs, K.O., and Stoker, J.J., "Forced Vibrations of Systems with Nonlinear Restoring Forces," *Quarterly of Applied Mathematics*, Vol. 1 (1943), pp. 97-115.

5. Friedrichs, K.O., *Advanced Ordinary Differential Equations*, New York, 1954.

6. Kac, A.M., "Forced Oscillations of a Nonlinear System with One Degree of Freedom Near to Conservative," *Prikl. Mat. i Meh.*, XIX (1955), pp. 13-32 (Russian).

7. Levinson, N., "Transformation Theory of Nonlinear Differential Equations of the Second Order," *Annals of Mathematics*, Vol. 45 (1944), pp. 723-737.

8. Loud, W.S., "Periodic Solutions of $x'' + cx' + g(x) = \varepsilon f(t)$," *Memoirs of the American Mathematical Society*, No. 31, 1959.

9. Loud, W.S., "Periodic Solutions of a Perturbed Autonomous System," *Annals of Mathematics*, Vol. 70 (1959), pp. 490-529.

10. Malkin, I.G., "Some Problems in the Theory of Nonlinear Oscillations," *A.E.C. Translation No. 3766*, Moscow, 1956.

11. Minorsky, N., *Nonlinear Mechanics*, Edwards Bros., 1947.

12. Poincaré, H., *Les Méthodes Nouvelles de la Mécanique Céleste*, 1892, 1893, 1899, reprinted by Dover Feb. 1957.